18979

507.8
GAR
Gardner, Robert, 1929-
Making and using scientific
models

L.A. MATHESON SECONDARY
LIBRARY

L.A. MATHESON SECONDARY
LIBRARY

MAKING AND USING SCIENTIFIC MODELS

ROBERT GARDNER AND ERIC KEMER

AN EXPERIMENTAL
SCIENCE BOOK

FRANKLIN WATTS
NEW YORK/CHICAGO/LONDON
TORONTO/SYDNEY

Diagrams by Vantage Art

Photographs copyright ©: Ford Motor Company: p. 6; American Small Business Computers: pp. 9, 13; Figg Engineers, Inc., Tallahassee, Fla.: p. 10; Boeing Corporation: p. 14 top; NASA: p. 14 bottom; Abbott Laboratories: p. 17; Henry Rasof: pp. 18, 19, 31, 151; Terence Dickinson: p. 44; Lick Observatory: p. 57; Stock Shop/Medichrome: pp. 67 (Richard Hirneisen), 74 bottom (A. Tsiaras); Tech Medica: p. 74 top; Photo Researchers, Inc.: pp. 90 (Clive Freeman, Royal Institution/SPL), 96 top (Dr. Mitsuo Ohtsuki/SPL), 96 bottom (Philippe Plailly/SPL), 122 (William Carter); International Salt Company: p. 100.

Library of Congress Cataloging-in-Publication Data

Gardner, Robert, 1929–
  Making and using scientific models / Robert Gardner and Eric Kemer.
      p.   cm.—(An Experimental science book)
  Includes bibliographical references and index.
  Summary: Discusses the use of scientific models to represent concepts in earth science, astronomy, biology, physics, and chemistry.
    ISBN 0-531-10986-0 (lib. bdg.)—ISBN 0-531-15662-1 (pbk.)
    1. Science—Exhibitions—Technique—Juvenile literature.
  2. Models and modelmaking—Juvenile literature. 3. Science—Mathematical models—Juvenile literature. [1. Science—Models.
  2. Models and modelmaking. 3. Science projects.]   I. Kemer, Eric.
  II. Title. III. Series.
  Q182.3.G36   1993
  507.8—dc20   92-21124 CIP AC

Copyright © 1993 by Robert Gardner and Eric Kemer
All rights reserved
Printed in the United States of America
5 4 3 2 1

CONTENTS

1
Models in Science and
Science Fairs
7

2
Models in Earth Science:
From Aquifers to Sunsets
21

3
Models in Astronomy:
From Eclipses to Double Stars
36

4
Models in Life Science:
From Organs to Genes
66

5
Models in Chemistry:
From Crystals to Chain Reactions
80

6
Models in the Physics of Motion:
From Falling Bodies to
Speeding Bullets
108

7
Models in the Physics of Light:
From Mirrors to Slinkies
120

8
Models in Engineering:
From Bridges to Airplanes
138

Bibliography
155

Index
157

MAKING AND USING SCIENTIFIC MODELS

Photo 1. A scale model of an automobile.

1
MODELS IN SCIENCE AND SCIENCE FAIRS

To simplify and help us better understand the complex world in which we live, we often build models to represent some aspect of nature. To most people, a model is a miniaturized version of a larger object. A model car, for example, looks like a real car scaled down by a factor of about 100. Of course, the scale model differs from the real car in ways other than its smaller size. It is often made from different materials, and its interior may lack many details. In fact, it almost never includes an engine with its associated transmission, cooling, suspension, and electrical systems. That is, the scale model does not function like a car. See Photo 1.

Despite being limited in detail and function, *scale models* can serve a practical use. They allow the designers of large and expensive manufactured objects to develop their ideas with greater ease, speed, and economy. Making scale models is a standard procedure in architecture, interior design, and landscaping, as well as in automobile, boat, and airplane design and in manufacturing. Artists, too, develop their ideas with scaled-down drawings and sculptures before committing them to large canvases or massive blocks of granite. Many types of designers now create their initial models on a computer. See Photos 2 and 3.

There is another class of models, sometimes called *analogues*, for which similarity in behavior is the primary purpose. Similarity in behavior may follow similarity in appearance. For example, analogues of large structures, such as bridges (see Photo 4), that are used to test the structural strength of the design, may be scale models as well. However, many other analogues used in science and engineering bear little resemblance to their real-life counterpart. In the decay of a sample of a radioactive element, a specific fraction of the element's atoms will decay in a fixed period of time, but it is impossible to predict which atoms will change. This process can be modeled by throwing a large number of dice. After each throw a certain fraction of the dice (about one sixth) will land with "6" facing up. In the model, these dice can represent decayed atoms, which are then removed before the next throw. Each throw of the dice can represent the specific period of time. Performing variations of this model experiment and comparing the results with actual radioactive decay experiments can provide some productive insights.

Another type of model that is very useful in science is a *mathematical model*. These models are not even physical objects; they are abstract ideas expressed in the form of mathematical equations. An example of a simple mathematical model is the following equation for the growth of the world's human population.

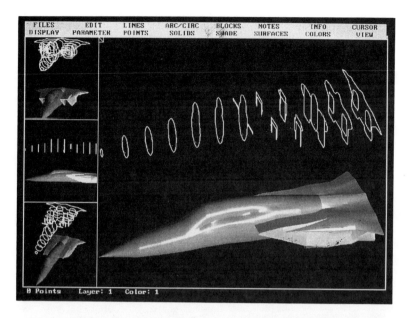

Photo 2. A computer-generated model of a jet

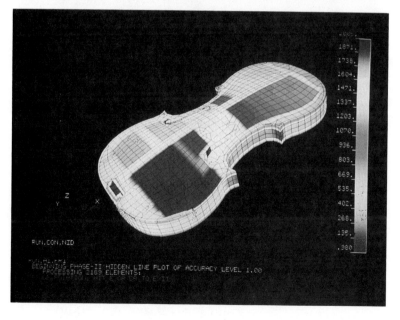

Photo 3. A computer-generated model of a violin

Photo 4. A model of a bridge

$$P = P_o (1.02)^n$$

In this equation P_o is the present population and P is the population n years from now. This mathematical model is based on the assumption that the population increases by a factor of 1.02, or 2 percent, each year. Often, mathematical models can be made much more visible by making a graph of the equation. Figure 1 is a graph of the equation above. Of course, such a model could not be used to predict the population for more than a few years. For example, this model predicts that 200

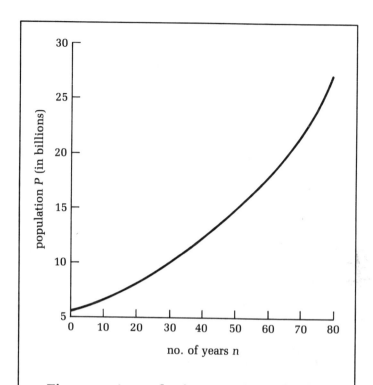

Figure 1. A graph of $P = P_0 (1.02)^n$. The projected population, P, is plotted on the vertical axis; the number of years, n, is plotted on the horizontal axis. As you can see, the population takes about 35 years to double if it increases at a rate of 2 percent each year.

years from now the world's population will be more than fifty times its current 5.5 billion. Most scientists believe that this is far more people than the earth can support. Any long-term predictions would require that the model be modified to include additional terms representing factors that might level out or lower the growth rate.

Mathematical models are perhaps the most common and useful of all models. They have found use in virtually every field of science and engineering. Any process that can be measured can be mathematically modeled with some degree of success. The growth of mathematical modeling has also been enhanced by the advent of powerful computers that can rapidly solve complex equations. With computers, scientists and engineers can develop, test, and refine mathematical models quickly, efficiently, and economically. See Photo 5.

Although mathematical modeling is widely used in science, processes remain that are still too complex to model even with the fastest and largest computers. One example is the weather. Despite efforts to model weather patterns with equations, meteorologists are unable to predict weather for more than a couple of days. And local weather patterns continue to provide surprises from one hour to the next.

Aerodynamics, which is the study of how complex shapes respond to the flow of air around them, is another example of a system that is too complex for mathematical models. In this field, scientists still use scale models in wind tunnels. In addition to scaling airplanes, cars, missiles, or other objects being investigated, scientists have learned to scale airspeeds and pressures in the wind tunnels to develop close similarity to real conditions. See Photos 6 and 7.

Scientists also use the term *model* to mean a theory—a carefully thought out system that explains some aspect of the natural world. For example, early scientists developed a theory or model to explain heat. The caloric theory, as it was

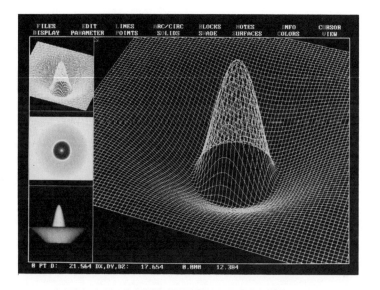

Photo 5. Mathematical equations and models can often be turned into graphic representations, as with this three-dimensional representation of a mathematical function. These functions have many applications in science and engineering.

called, assumed that heat was an invisible fluid attached to the particles of matter. The caloric fluid always flowed from warm bodies to cooler ones. Some substances, such as wood or coal, held an abundance of caloric, which explained why so much heat was released when they burned.

Despite their differences, all models share several important characteristics.

- They have some form of similarity—in appearance, behavior, or both—with the real aspect of the world they represent. The formal name for this property is *similitude*.

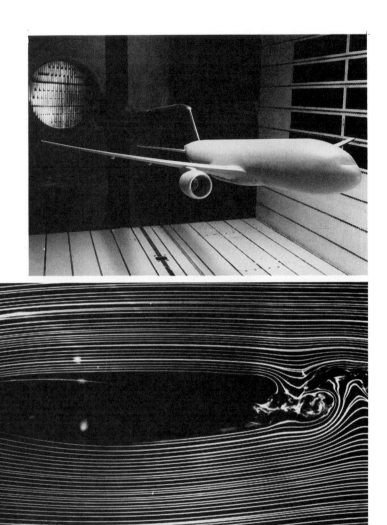

Photo 6. (Top) A scale model of the Boeing 777 is shown in one of the wind tunnels at a Boeing facility. The 777, which seats up to 400 passengers, will be in operation in 1995. Photo 7. (Bottom) Smoke flowing from left to right around a wing model in a wind tunnel at the NASA Langley Research Center. Note the turbulence at the rear of the wing.

- They are always simpler than the real objects they were designed to model. Any parts that are not essential in representing or understanding reality are eliminated from the model.
- They allow us to predict new relationships, to see things in ways that direct observation of reality may not allow.
- All models are developed with human brain power. They are a product of human observation and inventiveness. For this reason, it is important to keep in mind that a model is not reality itself but only a limited representation of it. Difficulties may arise if the limitations of a model are not recognized and the model is mistaken for reality. For example, you may see animated atoms on television with electrons orbiting the atom's nucleus. But in fact, we have no way of knowing what the path of an electron is. All we know is the probability of its position relative to the nucleus, which is only a tiny fraction of the atom's volume.

Because models are only attempts to explain or represent reality, we must constantly test the consequences of such models. This is done by carrying out experiments. If the results of an experiment agree with predictions made using the model, it gives us more confidence that the model is a good one. If experimental results do not agree with predictions made by the model, we know something is wrong with the model. It will have to be modified, or we will have to develop a new model that can explain the results of the experiment.

If you've ever visited a science museum, you know that models can be used to represent and simplify a great variety of scientific principles and processes. Furthermore, models of all kinds play a vital role in scientific thought and progress. In 1962, James Watson, Francis Crick, and Maurice Wilkins won a Nobel Prize for their discovery of the double-helix structure of the DNA molecule. Scientists in the 1950s had identified all the components of DNA (deoxyribonucleic acid), and they knew that the DNA in the chromosomes of living cells controlled the transmission of hereditary traits and the development of these traits in living organisms. However, they did not understand how these components fitted together until these three scientists devised their double-helix model of the DNA molecule. They even built a physical model of the molecule (see Photo 8).

MODELS AND SCIENCE FAIRS

Some models that you might build would be appropriate for a science fair. However, judges at such fairs do not reward models that are simply copies of drawings in books. Models of volcanoes, the solar system, and the human eye, so commonly found at these fairs, are often devoid of creative thought or imagination. If you decide to present a project involving a model, do something original and creative. A model airplane built by following a set of drawings may be fun to do if you enjoy working with your hands, but it will not win any prizes at a science fair. On the other hand, a wind tunnel used to test the efficiency of various wing structures that you have designed might be a big winner. Photos 9–12 show some models at a science fair.

Photo 8. Model of a DNA molecule

If you decide to enter a model in a science fair and have never done so before, be sure to read some books on science fairs such as those found in the bibliography at the back of this book. Such books will provide helpful hints and enable you to avoid the pitfalls that sometimes plague first-time entrants.

In this book we will ask you to make and use models of all types. We hope that in doing so, you will gain an appreciation for the powerful role that models play in science and engineering. We also hope that by building and working with models, whether for science fairs, class projects, or pure enjoyment, you will acquire a valuable skill that will serve you in your own future researches, whether they be scientific in nature or otherwise.

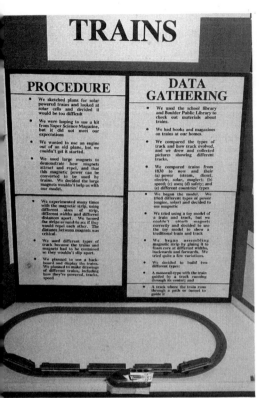

Photo 9. (Left) Train project exhibited at the 1992 Boulder (Colorado) Valley School District Science Fair

Photo 11. (Below) Model desert ecosystem at the 1992 Boulder (Colorado) Valley School District Science Fair

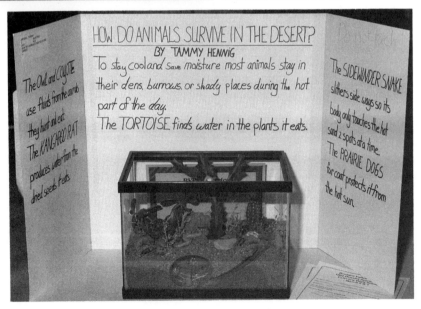

Photo 10. (Left) Solar car project at the 1992 Boulder (Colorado) Valley School District Science Fair

Photo 12. (Below) Model boat project at the 1992 Boulder (Colorado) Valley School District Science Fair

SAFETY FIRST

Most of the projects included in this book are perfectly safe. However, to ensure that you always work under safe conditions, read the following safety guidelines before you start any project.

1. Do any experiments or projects, whether from this book or of your own design, under the supervision of a science teacher or other knowledgeable adult.

2. Read all instructions carefully before proceeding with a project. If you have questions, check with your supervisor before going any further.

3. Maintain a serious attitude while conducting experiments. Fooling around can be dangerous to you and to others.

4. Wear approved safety goggles when you are doing anything that might cause injury to your eyes.

5. Do not eat or drink while experimenting.

6. Have safety equipment such as fire extinguishers, fire blankets, and first aid kits nearby while you are experimenting, and know where this equipment is.

7. Do not touch any high-voltage source or anything connected to a high-voltage source.

8. Never experiment with household electricity except under the supervision of a knowledgeable adult.

9. Light bulbs produce light but they also produce heat. Don't touch a lit high-wattage bulb.

10. Never look directly at the sun. It can cause permanent damage to your eyes.

2
MODELS IN EARTH SCIENCE: FROM AQUIFERS TO SUNSETS

If you compare the western coast of Africa and the eastern coast of South America on a globe or a map of the world, their shapes suggest that these two continents might once have been joined and that they have, over time, drifted apart. Careful modeling of plate tectonics along with precise measurements of movements in the earth's crust support the theory that these two continents were indeed joined millions of years ago. Plate tectonics is concerned with the movement of huge chunks of the earth's crust.

The natural processes involved in the study of the earth, such as weather, geological changes, earthquakes, the water cycle, and changes in the earth's crust are so large, complex, violent, or long term that it is difficult to draw conclusions from direct observation or to test ideas by direct experiment. Consequently, models play a vital role in our attempts to understand these processes.

While plate tectonics is a complicated subject, a number of other aspects of earth science lend themselves to somewhat less complicated modeling. You'll have an opportunity to examine and build some of these models in the sections that follow.

TOPOGRAPHIC MAPPING

Maps are among the most common and frequently used models. Their function is to represent certain aspects of a region's geography while excluding extraneous detail. By doing so, they provide their users with a means of understanding a geography that is too large or complex to view or grasp directly. Maps also provide a safe, efficient, and economical means of deciding and testing courses of action in advance.

The most common type of map is a simple road map, which represents the relative locations of roads and towns on a precise scale. A road map also represents types of roads (superhighway, toll road, one-way street) and shows such features as airports, parks, institutions, and major landmarks. Road maps generally omit details about the changing elevations and shape of the terrain. It is possible, however, to precisely model such features on a flat map. Such maps, known as *topographic maps*, are useful whenever details of an area's terrain need to be known. Military strategists, mountain climbers, highway engineers, architects, and landscapers are a few of the people who find such maps useful.

From Models to Maps

To gain an understanding of how topographic maps work, first mold a model terrain out of Plasticine clay on the bottom of a large wide-mouth jar or a clear plastic bucket. If you have a large amount of clay, you might consider using a small aquarium. For your first model, mold a single mountain. Then draw a vertical centimeter scale along the side of the jar starting from 0 cm at the bottom and ending at a level above the highest peak. Then pour water into the jar until it reaches the 1-cm level. Use a sharp pencil to gently trace the line where the water meets the clay. The line you have drawn is a contour line. Every point along this line represents the same elevation—1 cm in this case. If you were to photograph this clay model from above, you could identify every point where the elevation was 1 cm by looking along the contour line.

Continue adding water in 1-cm intervals and tracing the corresponding contour lines along your clay mountain. When you have finished, you will have a complete series of contour lines, each representing a unique elevation. To make a topographic map, transfer these contour lines to a flat sheet of paper and label them with their corresponding elevations. One way of doing this was already suggested—that is, by photographing the clay model from above. You can make a rough topographic map by letting your eyes serve as the camera. Can you think of more precise ways of making the transfer without using a camera?

For your second model, mold a terrain that includes two or more mountain peaks, a valley or canyon, and a gently sloping plane or plateau. Try to give the mountain peaks different heights and varying degrees of slope on each side. In other words, try to represent all the variety of form that might arise in the real world.

After you have drawn your map, compare the shapes and spacings of its contour lines with those of the clay model. What features on the map indicate steep slopes? Shallow slopes? Mountain peaks? Valleys?

From Maps to Models

As an additional exercise in topographic mapping, try to reconstruct a model terrain from the map shown in Figure 2. One way to do this is to first trace the closed loops formed by the contour lines onto separate pieces of cardboard. Then cut out each loop and stack them in their proper order with spacers in between. The spacers should be scaled appropriately. Finally, cover the resulting structure with a thin layer of papier-mâché.

- You might like to build three-dimensional scale models of some actual terrain using topographic maps obtained from your library, the United States Geological Survey, or a store that supplies goods for camping and mountain climbing. For example, imagine that you are about to lead a climbing expedition up Mount Everest and must chart the best course for ascent. Such a model might prove useful for planning as well as impressing your financial backers.

SOIL EROSION

The surface of the earth is constantly changing. Some of these changes, such as those produced by volcanoes and earthquakes, are sudden and violent. Other changes, such as those produced by continental drift and river erosion, require centu-

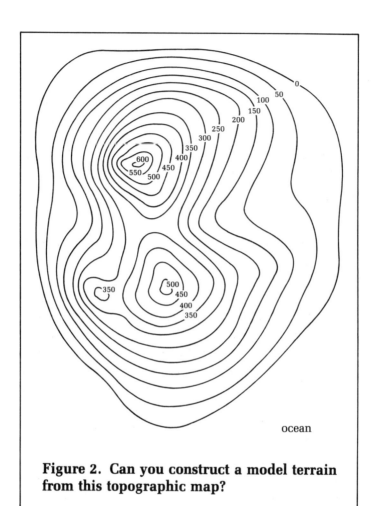

Figure 2. Can you construct a model terrain from this topographic map?

ries to produce small effects. Because of the vastness of the times, distances, and energies involved in geological changes, scientists have turned to models to help develop their understanding of such slow change.

One simple model that can illustrate the processes of stream formation and soil erosion can be made by using a wooden trough or a large tray from a nursery or greenhouse filled with gravel, sand, stones, topsoil, or various mixtures of these soils. Water sprayed from a hose or watering can plays the role of rainfall, which is the major agent of erosion.

This model can be used to investigate many factors that play crucial roles in large-scale erosion. For instance, the trough can be set at different angles and the water can be sprayed to represent rains that vary from a fine mist to a heavy downpour. Different proportions of gravel, sand, topsoil, and stones can be used to simulate different soils. Furthermore, the terrain inside the trough can be sculpted to form terraces, walls, and dams. Small plants can be used to model the effect of trees and other vegetation on erosion.

The number of experiments you can perform with this simple model is limited only by your imagination. But be sure to control the variables. Change only one variable at a time. For example, don't change both soil type (sand, gravel, stones, or topsoil) and rain (drizzle, steady, or downpour) at the same time. If you do, you won't know whether the changes are due to the type of soil or the kind of rainfall.

As you experiment, you'll be able to uncover many patterns of erosion. Even better, you may discover or invent novel ways to prevent erosion or landslides. However, the degree to which your findings and inventions apply to the real world will depend on your model's degree of similitude. First attempts at modeling rarely produce perfect similitude. You will find, as all scientists have, that models require constant questioning and refining.

DIGGING FOR TREASURE (AND FINDING WATER)

What happens to rain that falls onto the earth? Some of it evaporates quickly into the atmosphere or runs off into streams and rivers that flow into ponds, lakes, and oceans. Some rain is absorbed by plants. The rest seeps into the ground, giving rise to groundwater—the source of springs and wells.

Under the action of gravity, groundwater permeates downward through the soil until it reaches a level of saturation, that is, until water fills all the space between soil particles. The level that divides saturated from unsaturated soil is called the water table. Below the water table lies an aquifer, where the soil is saturated with water. You may have discovered the water table for yourself while digging for "buried treasure" as a child. In some regions—for example, the seashore—the water table lies only a meter or less below the surface. If you dig in such an area, you'll know when you reach the water table because water will seep into the hole.

The water table usually rises and falls with the contour of the land. However, the water table topography is generally less drastic than the land surface. This fact, which is illustrated in Figure 3, has two important consequences. First, the water table at the top of tall hills is not necessarily very far below the surface. It might be less than a meter deeper than it is at the bottom of the hill. Second, if the elevation of the land surface drops off rapidly, it can dip below the water table. This can result in the formation of ponds, lakes, and swamps.

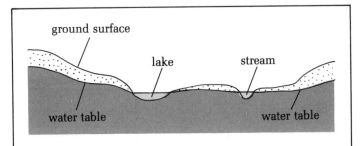

Figure 3. Often the topography of the water table is gentler than that of the surface.

Building a Model Aquifer

You can build a model water table and aquifer using soil, fine gravel or sand, and an aquarium, a clear plastic shoe box, or a large glass bowl. To make the model realistic, first cover the bottom of the aquarium with a somewhat irregular layer of Plasticine clay. The clay represents an impenetrable layer of bedrock beneath the soil. Then fill the aquarium with sand or soil and form some model hills, valleys, stream beds, and deep wells in it. The lowest soil elevation should be at least a few centimeters above the "bedrock."

Now, gently spray the surface with water or, better yet, colored water. Stop frequently and observe how the water seeps in. Where the soil and underlying clay meet, the glass or plastic wall should provide a good cross-sectional view of the soil.

Observe the formation of the water table. As more water is added and the water table rises, observe the formation of "swamps," "springs," and "ponds." To see how these might dry up during a drought, watch what happens if no water is added for several days. Then add more water and notice the level of the "aquifer" beneath "hills" and "valleys."

- Sometimes aquifers are stacked like plates, one above the other. Can you build a model of one aquifer above another?
- Can you build a model to illustrate how an artesian well is formed? An artesian well is a well drilled through rock in which water rises to the surface.
- What might cause water to move in an aquifer? How can you use your model to illustrate such movement?
- Use your model to show how an artificial pond could be constructed.

A WAY TO CLEAN WATER

Water stored in open reservoirs must be purified in several ways before it is fit for human consumption. The final step in this purification is the addition of chemicals that kill various germs. The earlier steps involve the removal of dirt particles, which is accomplished by filtration. In many reservoirs, the water is filtered by layers of sand and gravel that lie at the bottom. As the water permeates down through these layers under the action of gravity, the unwanted particles are trapped by sand. The water is then collected and carried off in porous pipes that lie beneath the sand.

Figure 4 is a diagram of a model reservoir filtering system that uses layers of fine sand, coarse sand, and pebbles as the filters. Build your own version of this system and test its ability to filter

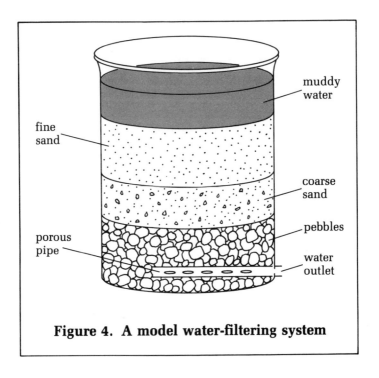

Figure 4. A model water-filtering system

muddy water. You can evaluate the purity of the filtered water by examining its clarity by eye or by using a microscope to look at drops of water that have passed through the filter. *Note: Do not drink the water.* Try modifying your model in various ways. For example, experiment with layers of different thicknesses or change the order of the layers in the filter. Grind up some wood charcoal, make it into a paste by adding a little water, and use it as the uppermost layer in the filter. Does the charcoal make the filter more effective? What other substances might you try in your model filter? Photo 13 shows a science fair project built around a model aquifer.

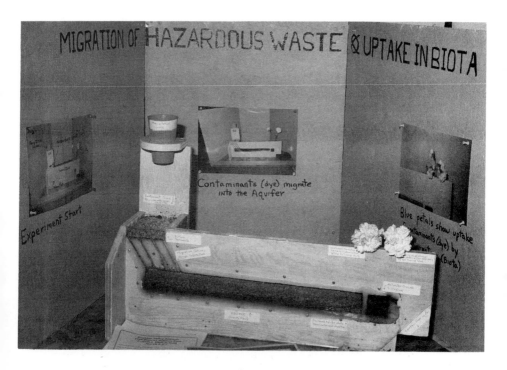

Photo 13. This model aquifer, entered in the 1992 Boulder Valley School District Science Fair, was used to investigate hazardous waste.

RAINBOWS

A rainbow is a common phenomenon that appears near the earth's surface. Perhaps you've noticed that a rainbow appears when sunlight strikes falling raindrops. The physics of a rainbow is quite complicated, but you can make a somewhat simplified model by using a shallow pan of water to represent falling raindrops. A mirror is used to produce the reflection that normally occurs within the raindrops.

A setup of the model is shown in Figure 5. Support one end of the mirror so that it makes a small angle to sunlight coming through a window or door. When light passes from air to water it is bent, or refracted. This refraction causes the light from the sun to separate into different colors because the different colors of light are refracted by different amounts. By adjusting the mirror, you can see these colors on a white ceiling or wall, or on a paper screen that you can hold above the refracted beam of sunlight that has passed through the water and been reflected.

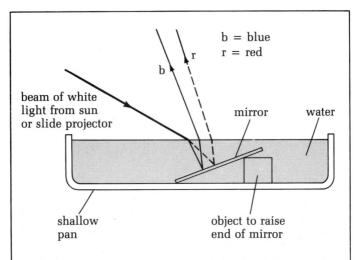

Figure 5. Light with short wavelengths, such as blue light, is refracted (bent) more than light of longer wavelengths, such as red light, when it passes from air into water or from a vacuum into air. Therefore, sunlight, which is made up of both long and short wavelengths, can be separated into light of different colors.

- Are the colors that you see here in the same order, top to bottom, as they are in a natural rainbow? If not, can you explain why the order is different?
- On a warm, sunny day, spray water from a garden hose into the sunlight to make a model of a natural rainbow. Where should observers stand to see the rainbow?

BLUE SKIES AND RED SUNSETS

As you saw in the previous section, blue light is bent more than the other colors when white light is refracted as it passes from air to water. The same would hold true if the light were to pass from air into glass. That observation is useful in developing a model to explain why skies are blue and sunsets red. It shows you that different colors of light may behave somewhat differently.

When light passes through the earth's atmosphere it is scattered; that is, the light energy is absorbed by molecules of air and dust and then reemitted. The light released by these particles in air travels in all directions, not just in the direction it was originally moving. Blue light, which you've seen is refracted more than red light, is also scattered much more than red light by the earth's atmosphere. As a result, the scattered light coming to our eyes from molecules and particles in the sky above our heads is bluish. For the same reason, the sun has a yellowish color. Light from the sun is actually white (a mixture of all the colors of light), but much of its blue light is scattered by the atmosphere leaving a mixture that appears yellow when it reaches our eyes.

When the sun is near the horizon, its light

passes through a much greater length of the atmosphere, as you can see from Figure 6. Therefore, even more blue light is scattered, causing the sun's color to slowly change from yellow to red as the sun approaches the horizon.

Although molecules of air are far apart in comparison with molecules in a liquid, the atmosphere is many kilometers thick. As a result, sunlight encounters a vast number of molecules before it reaches the earth. You can take advantage of the

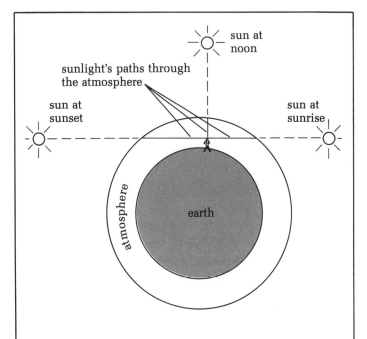

Figure 6. Light from the sun passes through a greater length of atmosphere at sunrise and sunset than at other times of the day, which means it strikes more air molecules at the beginning and end of each day.

greater concentration of molecules in a liquid to make a model of the scattering of light by the atmosphere. A fish tank filled with water in a dark room can serve as the "atmosphere."

Place a slide made of black film with a round hole cut in the center into a slide projector. Place the lens of the projector against one end of the tank. The round beam of light from the projector represents sunlight. To enhance the scattering, you can add dust particles to the "atmosphere" in the tank. Drops of milk or small amounts of a powdered nondairy creamer work well. Simply stir the milk or powder into the water.

Notice what happens to the color of the "sun," seen through the opposite end of the tank from the projector, as you add more and more milk or nondairy powder. Notice also the color of the "sky," seen from the side of the tank, as the scattering of the light increases.

- Can you design other models for the scattering of light?
- Devise a model to show how scattering affects the passage of light through fog. Can your model be used to explain why fog lights are yellow?
- You may have heard the phrase "once in a blue moon" used to indicate a very rare happening. But blue moons do occur. In Edinburgh, Scotland, in September 1950 a blue moon was observed at night and a blue sun at sunset. Can you find any molecules that scatter more red light than blue light? Can you produce a model for making a blue moon?

3
MODELS IN ASTRONOMY: FROM ECLIPSES TO DOUBLE STARS

Our ideas about the universe—our "models" of the universe—have been changing for thousands of years. Originally these models were based on naked-eye observations and a lot of speculation. As science began to take shape and instruments, such as telescopes, were invented for viewing the heavens, these models changed. Today, with many sophisticated instruments, including the Galileo Space Telescope, our models have grown more sophisticated and our speculation has lessened.

Still, except for a foray to the moon, all we know about the universe is based on what we learn from different types of radiation from these heavenly bodies—visible light, X rays, microwaves, etc. Unlike other scientists, astronomers must rely upon indirect contact with the objects they investigate. They cannot touch a star, step on a distant planet, or fly to another galaxy.

While the instruments astronomers use to collect data are often millions of light-years from the sources of radiation to which they respond, they provide the evidence that leads us to rebuild and modify our conceptual models of the universe and the celestial objects within it. For example, recent data collected from the heavens suggest that dying stars more than three times as massive as our sun may produce black holes, regions in space where gravity is so intense that nothing, not even light, can escape.

Note. In some of the activities in this chapter you will be asked to look close to the sun. Do this very carefully and **be careful never to look directly at the sun. Doing so can cause permanent damage to your eyes.**

THE PHASES OF THE MOON

Anyone who watches the moon, even casually, knows that its appearance changes dramatically during the course of a month. After the moon disappears for a day or two, we first see it again as a thin crescent close to the setting sun. As days pass, the apparent distance between the sun and moon increases. If you align one arm with the setting sun and the other with the moon, you will see that the angle between moon and sun also increases as they move apart. The angle between moon and sun increases each day for about a week until we see the moon at first quarter looking like a disk that has been cut in half. A week later, we see a full moon rising in the east as the sun sets along the western horizon. After this, the moon's disk grows smaller with each passing day. About a week after full moon, it reaches third quarter. Then we see the opposite half of the disk we saw at first quarter.

The moon continues to "shrink" until we again see a crescent moon, the reverse of what we saw earlier, rising at dawn and leading the sun upward from the eastern horizon.

Modeling the Moon's Phases

Assume that the moon moves in a nearly circular orbit about the earth. To make a simple model of the moon's motion, let your own head represent the earth. Let a white tennis ball resting on the tips of your fingers be the moon. If you turn with your hand outstretched, you have a model of the moon's orbit about the earth. Now add a sun to your model by placing a lamp with a glowing incandescent bulb on one side of an otherwise dark room. If you stand on the opposite side of the room and observe the tennis ball moon moving slowly around your head, you'll be able to see the various phases of the moon.

Where is the moon relative to the earth and sun when it is full? When it is at first quarter? At third or last quarter? When it is a crescent at sunset? A crescent at sunrise? When it is a new moon (invisible)? In which phase must the moon be if there is to be a lunar eclipse? A solar eclipse?

HOW FAR AWAY IS THE SUN?

Using a geometric model—a mathematical model that involves points, lines, angles, and surfaces—you can estimate the distance to the sun if you know the distance to the moon. The distance to the moon can be determined by simultaneously establishing sight lines to the same position on the moon from points on the earth that are far apart. The method is shown in Figure 7.

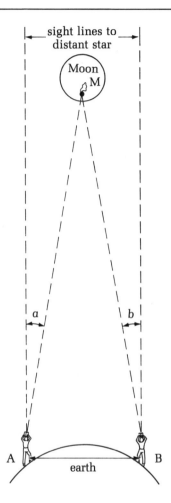

Figure 7. Two observers at distant points A and B on the earth can measure the distance to the moon. By knowing the distance from A to B and angles a and b, which are measured from the parallel sight lines to a distant star, the distance to M (a point on the moon) can be determined.

Both observers, A and B, measure the angle between a point on the moon and a distant star that is nearly in line with the moon. Since the star is very far away, we can assume that the two sight lines to the star are parallel. The distance between the two observers, AB, can be measured. By making a scale drawing of the triangle ABM shown in Figure 7, you can measure the triangle's altitude and estimate the distance to the moon. The results of such an experiment indicate that the moon is about 380,000 km from the earth.

With a good value for the distance from the earth to the moon, you can now make a rough estimate of the distance to the sun using a geometric model. If you made a model of the moon's phases in the previous section, you know that when the moon appears as exactly half a disk, the angle between the imaginary lines connecting the moon to the sun and to the earth is 90°. See Figure 8a.

You can confirm this in a rough way by extending your arms and measuring the number of fists between the moon and the sun. **Do not look directly at the sun. It can damage your eyes.** You'll find they are separated by just about nine fists. A fist at the end of an extended arm covers just about 10°.

To see that this is true, extend your left arm straight out in front of you so that the top of your fist is in line with the horizon. Then place your right fist on top of your left and go fist over fist upward. How many fists does it take before you are directly above your head?

Knowing that the angle from earth to moon to sun is 90°, you need to measure the angle from moon to earth to sun in order to make a scale drawing and estimate the distance from earth to sun. You can find this angle in the following way.

1. Stick two pins in a piece of paper resting on a sheet of cardboard to make a sight line to the moon as shown in Figure 8b.

2. Use a ruler and pencil to draw the straight line between the pins. Extend the line along the entire width of the paper.

3. Place a tall pin or nail to one side of the line.

4. Ask a friend to keep the two sight pins in line with the moon while you make a mark at the end of the shadow cast by the pin as shown in Figure 8c.

5. Use a ruler to extend the line along the pin's shadow until it meets the sight line to the moon. (How do you know that the line established by the shadow lies along the line that connects the earth and the sun?)

6. Use a protractor to measure angle a, the angle between moon, earth, and sun. How large is this angle? What does your measurement of angle a tell you about the distance from earth to sun as compared with the distance from earth to moon?

7. Using the 90° earth-moon-sun angle and the angle you have just measured, make a scale drawing of the triangle made by the earth, moon, and sun. Remember, the short side of the triangle represents the earth-to-moon distance, which is 380,000 km. How far away is the sun according to your drawing? How accurate do you think is your determination of the distance to the sun?

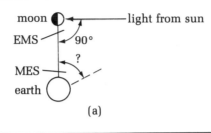

(a)

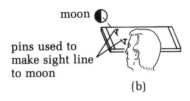

(b)

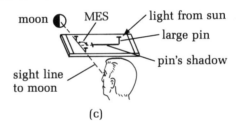

(c)

Figure 8. You can make a scale drawing to estimate the distance from the earth to the sun. (a) When the moon is at first quarter, angle EMS is 90°. (b) To find angle MES make a sight line to the moon using two pins. (c) The shadow of a large pin, which lies along the line from sun to earth, can be extended until it hits the sight line to the moon. With these two lines we can measure the angle between the sun, earth, and moon (angle MES in the drawing). Knowing the angles EMS and MES and the distance to the moon, you can complete the task.

- Suppose your measurement of angle a had been off by half a degree. What effect would this have had on your determination of the distance from the earth to the sun?
- Can you design a better way to measure the distance to the sun?

A NEW MODEL OF THE UNIVERSE

Early Greek astronomers believed that all celestial bodies moved in perfect circles while attached to invisible concentric spheres of different radii, all with centers coinciding with the center of the earth. It was difficult not to agree with this model, which placed the earth at the center of the universe, because that's certainly what you observe if you look at the heavens.

Then, in 1543, the Polish astronomer Nicholas Copernicus (1473–1543) proposed a new model of the universe in his book *On the Revolutions of the Celestial Spheres*. The Copernican model was sun centered, or heliocentric (*helio* is Greek for "sun"), rather than earth centered, or geocentric (*geo* is Greek for "earth"). According to Copernicus, the apparent daily movement of the sun, moon, planets, and stars about the earth was caused by the earth's rotation. The changes in the positions of the stars resulted from the earth's movement in its yearlong orbit about the sun. See Photo 14.

The Copernican model enabled astronomers to measure the distance between the sun and the planets, at least in terms of the distance between the earth and the sun. The distance from Earth to the sun was defined as 1 astronomical unit (1 AU). Using the astronomical unit as their basis of mea-

Photo 14. Star trails

surement, astronomers soon found the radii of the orbits of the known planets. These radii are found in Table 1. Notice that Uranus, Neptune, and Pluto had not been discovered in the sixteenth century.

Table 1. Radii of the orbits of the planets in AU

Planet	Radius of orbit
Mercury	0.39
Venus	0.72
Earth	1.00
Mars	1.53
Jupiter	5.20
Saturn	9.54

EXPLORING THE SOLAR SYSTEM
BODES WELL

Look closely at the data in Table 1. Do you see any pattern in the numbers found there?

Before you give up on finding a pattern, try this.

1. Write down the series of numbers 0, 3, 6, 12, 24, 48, 96, ... As you can see, after the second number, each number is simply twice the preceding one.

2. Add 4 to each number in the sequence. What is the new sequence of numbers?

3. Divide each number in the new sequence by 10. What is each number in this last sequence?

Compare the numbers obtained in the last sequence with the radii of the planets' orbits in Table 1. What do you notice? Is this a coincidence, or does it have some significance?

In the late 1700s, many years after the radii of the planets' orbits were known, a mathematical model or rule was made popular by a young German astronomer, Johann Bode (1747–1826). The

[45]

rule, which is the pattern you discovered, came to be known as Bode's law. You'll notice that there is no planet with an orbit of 2.8 AU. Was there an undiscovered planet at this distance from the sun?

If you do some research at your school or local library, you'll find that there is something—or there are some things—2.8 AU from the sun, but not a planet. What is it or what are they?

- Does Bode's law hold for the planets discovered later—Uranus, Neptune, and Pluto? Their distances from the sun were found to be 19.18, 30.06, and 39.44 AU, respectively.

Table 2 contains the periods and masses of the planets, relative to Earth, as well as the radii of their orbits. In addition to Bode's Law, perhaps you can find other mathematical models that work for the planets. When the German mathematician and astronomer Johann Kepler (1571–1630) compared the ratio of the cube of the radius of each

Table 2. Data about the planets

Planet	Radius of of orbit (AU)	Period of revolution (years)	Mass (Earth = 1)	Radius of planet (Earth = 1)
Mercury	0.39	0.24	0.05	0.40
Venus	0.72	0.62	0.81	0.99
Earth	1.00	1.00	1.00	1.00
Mars	1.53	1.88	0.11	0.54
Jupiter	5.20	11.86	317.73	11.25
Saturn	9.54	29.46	94.82	9.45
Uranus	19.18	84.17	14.72	4.18
Neptune	30.06	164.56	17.22	3.89
Pluto	39.44	247.47	0.01	0.5

planet's orbit to the square of its period of revolution, that is, R^3/T^2 he found an interesting relationship, which later became known as Kepler's third law. What was this law?

HOW FAR OUT ARE THE SUN AND PLANETS?

Today, we can determine the distances to the sun and planets very accurately. We send a burst of radar waves and wait for the echo to return. Since radar waves travel at the speed of light (300,000 km/s), we can easily find the distance by noting the time delay between sending the signal and receiving the echo. For example, a radar signal sent to Venus returns in about 280 seconds, or 4.7 minutes. In 280 s, a radar signal will travel 84,000,000 km (300,000 km/s x 280 s). Of course, the signal has to travel to Venus and back; therefore, the actual distance to Venus is half of 84,000,000 km, or 42,000,000 km.

Actually, astronomers knew these distances quite accurately as long ago as the late seventeenth century. In 1672 Giovanni Cassini in Paris and his colleague, Jean Richer, in the colony of Cayenne in French Guiana, about 8,000 km away, simultaneously measured the angle between Mars and a distant star as shown in Figure 9. The angular difference, which was less than 0.006°, was measured with accurate instruments. The known distance between Paris and Cayenne together with the angle θ revealed that Mars, which is about 0.52 AU from Earth, is also about 78,000,000 km from Earth. From Cassini and Richer's measurements, it follows that 1 AU is equivalent to 150,000,000 km. It was then a simple matter to convert astronomi-

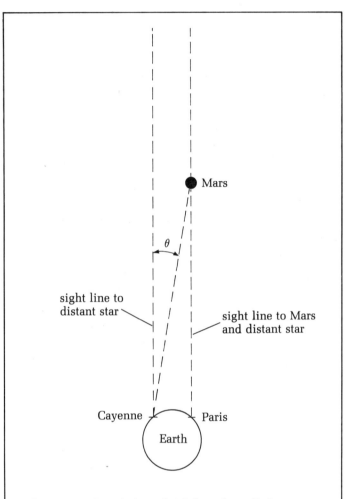

Figure 9. Cassini and Richer found the distance from Earth to Mars. Both of them looked at a very distant star and Mars at the same time. In Paris, Cassini saw Mars and the distant star along the same line of sight. Richer, in Cayenne, saw a small angle, θ (greatly exaggerated here), between Mars and the same star.

cal units to kilometers or miles and find the actual distances to the other planets.

FINDING THE DIAMETER OF THE SUN

In theory, since the distance to the sun is known, one could measure the angle across the sun, which is about half a degree, and then make a scale drawing to determine the sun's diameter. **Such a method is not appropriate for naked-eye observations, however, because looking directly at the sun can cause blindness.**

An approach that you can use safely is to make a pinhole image of the sun. Near the closed end of a large box, such as the kind movers use to hang clothes in, cut a small, square opening about 2 cm on a side. Tape a piece of black construction paper over the opening. Then use a pin to punch a small hole in the center of the black patch. Hold this large box over your head as shown in Figure 10. The side that has the pinhole should be above and behind your head. On the other side of the box, directly opposite the pinhole, tape a sheet of white paper. Stand with your back to the sun and your head just below the pinhole. Turn the box until a pinhole image of the sun appears on the white paper. Use a pen or pencil to carefully mark the right and left edges of the image.

Now measure the distance, D, between the pinhole and the screen. You can then remove the paper and measure the diameter, d, of the sun's image. From the geometry of Figure 11, you can see the two triangles, T and T^1, are similar. You know D, the distance between pinhole and screen; and you know d, the diameter of the sun's image. You also know that the distance to the sun, D^1, is

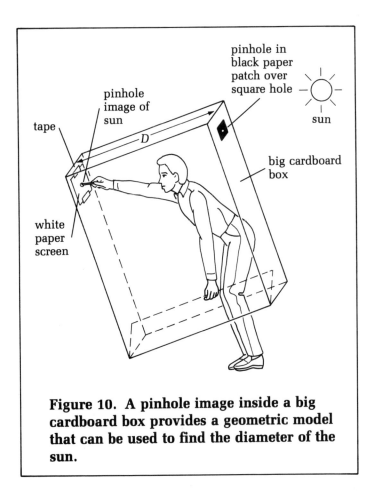

Figure 10. A pinhole image inside a big cardboard box provides a geometric model that can be used to find the diameter of the sun.

1 AU, or 150,000,000 km. The only unknown is d^1, the diameter of the sun. But

$$d^1 = D^1 \times \frac{d}{D} \text{ or } 150{,}000{,}000 \text{ km} \times \frac{d}{D}.$$

What do you find the diameter of the sun to be using this geometric model?

[50]

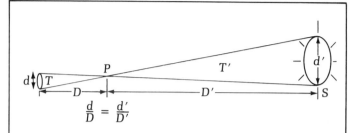

Figure 11. In this geometric model of the sun and its pinhole image, *d* is the diameter of the pinhole image of the sun, *D* is the distance from the pinhole at *P* to the image, *d'* is the diameter of the sun, and *D'* is the distance from the pinhole (or earth) to the sun, *S*.

HOW WIDE IS THE MOON?

It's safe to look at the moon, and you can use another geometric model to measure its diameter, given that the moon is about 380,000 km from Earth.

Carefully place your eye just above the end of a stick that you can hold along a sight line to the moon. Have someone move a coin along the stick until it just blocks out the moon or until its curvature matches the moon's curvature. Measure the diameter of the coin and the distance from the coin to your eye.

Using the information you have collected and the distance to the moon, construct a geometric model that will enable you to measure the diameter of the moon. What do you find the diameter of the moon to be? How does your value for the moon's diameter compare with the value that you can find in an astronomy book or encyclopedia?

HAVE GEOMETRY, WILL MEASURE (THE SIZE OF THE EARTH)

Eratosthenes, a Greek astronomer who lived from 276 to 194 B.C., was probably the first person to use a geometric model to measure the size of the earth. Eratosthenes lived in Alexandria, Egypt, just 800 km north of the city of Syene. In Syene on June 21 of each year the sun's image could be seen in a deep well at midday. At that same time, vertical objects in Syene cast no shadow. Eratosthenes knew that the sun would cast no shadows only if it were directly overhead. He knew too that this day (June 21) marked the summer solstice, the day when the sun would reach its greatest altitude above the horizon and its most northern path across the sky.

Eratosthenes's model of the earth was similar to that of most educated Greeks of his day. He believed that the earth was a sphere. He had probably seen the earth's curved shadow pass across the moon during a lunar eclipse, and he had seen the sails of a ship moving out to sea disappear slowly over the horizon. He knew that the earth was large, but he didn't know how large.

As he thought about the sun above Syene, he realized that he might use a geometric model to calculate the earth's circumference and thereby its diameter. Since Syene was 800 km south of Alexandria, objects would cast shadows at midday in Alexandria. In fact, on this day Eratosthenes observed that a tower 15.4 m tall cast a shadow just 2 m long.

With these measurements in mind, he probably drew a picture something like the one in Figure 12a. He realized that the vertical tower, its shadow, and the sun's rays that just skimmed the edge of the tower's top made a triangle. Next, he probably drew the triangle with the tower as the vertical leg and its shadow as the horizontal leg as shown in Figure 12b. He could then measure angle α and find that it was 82.6°.

Once he had drawn the triangle and measured the angle, he realized that in moving 800 km along the earth, its curvature had changed the angle between the sun's rays and the earth's radius by 7.4°, from 90° to 82.6°. He may also have drawn triangles like WAC and AST in Figure 12a and realized that they are similar. (The curvature in such small arcs is very small, so the figures can be considered to be triangles.)

- Using the globe you used in the previous section, make a model of Eratosthenes's experiment. A toothpick can represent the tower. A short length of soda straw can be used to find the "well" where the sun is overhead. How does the tower's shadow change as it is farther north or south of the well directly beneath the sun?

- Using the information gathered by Eratosthenes, you can make an estimate of the earth's circumference. What is your estimate? Based on your value of the earth's circumference, what is its diameter? How close is your estimate to the actual diameter of the earth, which is nearly 12,800 km? What was your percentage of error?

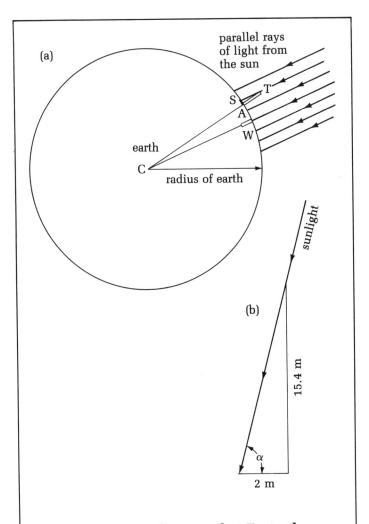

Figure 12. (a) A diagram that Eratosthenes might have made when he calculated the earth's diameter. W is the well at Syene, AT is the height of the tower at Alexandria, and AS is the length of the tower's shadow.
(b) The triangle made by the tower, its shadow, and sunlight.

EARTH AND MOON TO SCALE

Books on astronomy often contain drawings of the earth and its moon. Do you think such drawings are made to the proper scale? You can answer such a question on the basis of the work you did using geometric models earlier in this chapter.

- From the information you have gathered, what is the ratio of the diameter of the earth to the diameter of the moon? Since the moon's center is about 380,000 km from the earth's center, how does the distance between the earth and the moon compare with the diameter of the earth?
- If you use a ball with a diameter of 2.5 cm to represent the earth, what size ball should you use to reasonably represent the moon? Mount the "earth" on a vertical pin fastened to one end of a meter stick or yardstick. To make a scale model of the earth and moon, where on the stick should you fasten the ball used to represent the moon? Once you have determined the proper position, mount the "moon" on the stick in the same way you did the "earth."
- The sun is very large relative to the earth and moon. If you were to include the "sun" in your scale model, how large would it be? Where would you place it? Why is it not practical to include the sun in this earth-moon scale model?
- Have someone hold the stick with the scale model of the earth and moon in sunlight. **Do not look at the sun. Doing so can cause serious damage to your eyes.** With a sheet of

white cardboard held several meters from the stick, you can "capture" the shadow of the "earth" or "moon." Is the shadow sharp or fuzzy? What happens to the size of the darker part of the shadow as you move the screen closer to the stick? What happens to the size of the fuzzy part of the shadow? Can you explain why? Can you move the screen so far away that the dark portion of the shadow, which is called the umbra, disappears, leaving only the lighter, fuzzy shadow called the penumbra? Make a pencil-and-paper model to show why the shadow cast by the ball in sunlight has an umbra where no light can reach, and a penumbra where the shadow is partially illuminated.

ECLIPSING THE SUN AND MOON

A lunar eclipse—an eclipse of the moon—occurs when the shadow cast by the earth in sunlight falls on the moon. A solar eclipse occurs when the shadow cast by the moon in sunlight falls on the earth. See Photo 15.

Using your scale model of the earth and moon, make a model of a lunar eclipse. Then make a model of a solar eclipse using the same model of the earth and moon. **Do not look at the sun. It can cause serious damage to your eyes.**

After making eclipses using your model, can you explain why solar and lunar eclipses do not occur monthly, that is, every time the moon orbits the earth? Using your model in sunlight, look closely at the shadow the earth casts on the moon during a lunar eclipse. What evidence did early observers have for believing the earth was round?

Photo 15. *A total eclipse of the sun*

- During an eclipse of the sun, the umbra of the moon's shadow covers only a small area of the earth. Only inside the umbra is the eclipse of the sun total; that is, only within the umbra is the entire sun hidden from view. Make a line drawing—a two-dimensional model—like the one in Figure 13 to show the formation of the umbra and penumbra during a solar eclipse. Use the drawing to explain what observers in the larger penumbra area of a solar eclipse will see.
- The area of the umbra that touches the earth is very small; only the tip of the cone-shaped umbra touches the earth. If you know the distances to the moon and sun, as

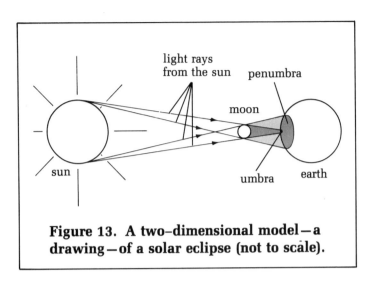

Figure 13. A two-dimensional model—a drawing—of a solar eclipse (not to scale).

well as the moon's diameter, can you find the approximate diameter of the sun from the two-dimensional model you have drawn? What do you find the diameter of the sun to be according to your calculation using this model? How does it compare with the diameter you found by making a pinhole image of the sun?

By using different models you can check one indirect measurement against another for the same object. Repeated agreement lends confidence to your findings. A single disagreement, however, if confirmed by others, will lead to careful reviews of earlier calculations and models that led to the previous measurements.

In what ways is your earth-moon system on a stick a reasonable model? In what respects does it lack agreement with reality?

A VIEW OF EARTH FROM SPACE

If you've seen photographs of the earth taken from the moon or from satellites far above the earth, you know that the world looks very different from afar. Using a globe, sunlight, and a knowledge of the position of Polaris, the North Star, you can make a model of the earth as seen from space—a model that will help you to understand some of the seasonal changes that take place on our globe.

Place a globe on a large empty can in bright sunlight. (If the globe is attached to a supporting stand, you'll have to remove the stand.) Orient the globe so that the place where you live is at the top of the globe. Then turn the globe so that its North Pole is pointed toward the North Star. If you don't know the altitude and location of the North Star, you can find it on a clear night, orient the globe, and then cover the globe with a sheet of plastic to protect it.

Once you have the globe properly oriented in sunlight, fix a vertical pin to the top of the globe with a piece of clay. The pin can represent a flagpole in your town. Compare the direction of the pin's shadow on the globe with the direction of your own shadow on the ground. How do the directions of these shadows compare?

Stand back and view your model. What you see is the earth as it would appear to someone in a spaceship far from the earth. What parts of the earth are now dark? Along what line is the sun now rising? Setting? What time of day or night is it in various parts of the world right now?

- An object directly beneath the sun when it is overhead will not cast a shadow. Run a pin through a short length of soda straw. The pin will serve as a handle so that you can move the soda straw along the globe. Find the place on earth where the sun is now directly overhead. Where is it? If you start this activity shortly after sunrise and continue until just before sunset, you can map the sun's path along the earth today. By repeating this process every month for a year, you'll have a model that reveals the way the sun's path across the earth changes from season to season. What is the sun's northernmost path along the earth? It's southernmost path? When do you find the sun moving along the earth's equator?

- During the year there are times when some places on the earth are dark all the time. At these times there are other places where it's daylight for all twenty-four hours. Use your model to find out at what times of the year this occurs. Where is it dark for months at a time? Where is it light for months at a time? At what times of the year does this take place? When does the sun rise exactly in the east and set directly in the west? Where does the sun rise and set the rest of the time?

- You can add another dimension to your model by using a tennis ball to represent the moon. Where should you place the "moon" so that it represents a full moon? A new moon? Other phases of the moon? Will a moon that appears to be full to you be seen as a full moon by everyone on earth, or just by people in your area?

- Bring the globe indoors to a large room that has a bright light at its center. Using the walls of the room as the earth's orbit, make a model that shows how the seasons change as the earth moves along its orbit around the sun. Modify the model to show the sun's path along the zodiac.

TO SCALE OR NOT TO SCALE

Figure 14a is a diagram of a satellite in orbit about the earth. Figure 14b represents the eccentricity of the earth's orbit—the amount it differs from a circular orbit. Drawings similar to these are often seen in earth science books. Do such drawings reflect the proper scale of satellites' orbits relative to the earth's diameter? The proper scale of the earth's own orbit? Are relief globes of the earth made to scale?—that is, are the heights of the mountains on the globe in the same ratio to the radius of the globe as the actual heights of the mountains are to the radius of the earth?

Using the information given in Table 3, you can answer these questions by making your own drawings of orbits and mountains heights drawn to scale. What do you find?

A SOLAR SYSTEM OF YOUR OWN

From your work in earlier sections of this chapter, you know the diameter of the earth, moon, and sun and the value of an astronomical unit in kilometers. Now that you've made several scale models and tested some others, you might like to build a scale model of the solar system.

To build such a model, you will need the information in Table 2 as well as your own measurement of the diameters of the sun and earth. You

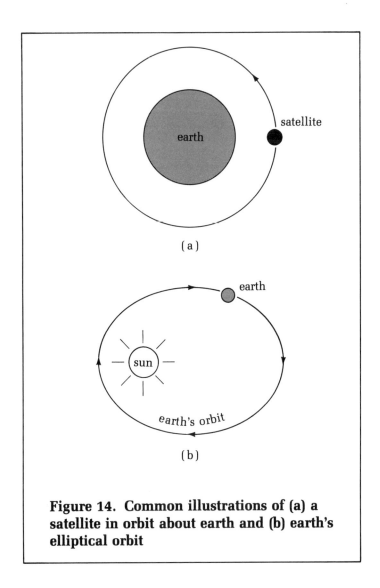

Figure 14. Common illustrations of (a) a satellite in orbit about earth and (b) earth's elliptical orbit

can choose spheres of the appropriate size to represent the sun and planets. For example, if you use a basketball, which has a diameter of 24 cm, to represent the sun, you'll need a ball bearing about 2.2 mm in diameter to represent the earth. Perhaps

Table 3. Earth and satellite data

Earth's perihelion (closest distance from sun) = 147,000,000 km
Earth's aphelion (farthest distance from sun) = 152,000,000 km
Earth's radius = 6,380 km
Altitude range of many nearby earth satellites = 240 km to 1,000 km
Altitude of communications satellites (geosynchronous orbit) = 35,500 km
Altitude of Mount McKinley = 6,194 m, or 6.194 km above sea level
Altitude of Mount Everest = 8,848 m, or 8.848 km above sea level.

you'd rather use a beach ball to represent the sun. What can you use then to represent the earth?

In any case, determine the size of the sun and planets and their separations before you start building your scale model. You'll be surprised at how much space will be needed.

- Can you build a dynamic model of the solar system in which you use elastic, magnetic, or some other force to represent the gravitational attraction between each planet and the sun? You may prefer not to use the actual scale between planets and sun in preparing this model, but you might want to make the periods of revolution (see Table 2) agree with reality.

THE EXPANDING UNIVERSE

For centuries, astronomers thought that the universe was static, or unchanging. They believed that it had always looked as it does now. But since the 1930s careful examination of the light coming from distant stars and galaxies, as well as an abundance of other evidence, indicates that the universe is expanding. The evidence for a growing universe—evidence that you might like to read about in various books on astronomy—has led scientists to develop a more dynamic model of the universe. Today, most astronomers believe that the universe came into existence about 20 billion years ago with a giant explosion that caused the matter making up the early universe to be blown apart. The Big Bang theory, as this model is called, holds that the universe has been expanding ever since that original explosion.

One concrete way of thinking about such a universe is to consider the "fabric of space" itself to be like the surface of a balloon. In this model, space is given only two dimensions, one less than the real space you're familiar with, but it also has curvature. Although the idea of curved space is a strange concept, Einstein's general theory of relativity asserts that space is curved. The curvature is the result of the presence of mass. The greater the concentration of mass in a given region, the more the space is curved around that mass. You read earlier about black holes, collapsed stars believed to bend the space around them so much that even light cannot escape.

It is not known for certain that the overall curvature of the universe is like the closed surface of a balloon. However, if space is closed in this way, then it would be possible (in theory) to take off in a spaceship, travel in one direction, and wind up in the place you started, just as you can circle the earth.

Making a Two-Dimensional Model of the Universe

To make the two-dimensional balloon model of the universe more visual, blow up a round balloon until it is about the size of an orange. Then use a marking pen to make a series of randomly placed dots all over the balloon's surface. Each dot represents a galaxy. Hold the balloon in front of a mirror and slowly but steadily blow more air into it. You can see that the distance between the dots increases. This is similar to what astronomers have observed is happening to the distances between the galaxies that make up the universe. What do you think happens to the size of the galaxies themselves as the universe grows? Test your hypothesis.

Pick any one dot to represent the Milky Way, the galaxy we live in. Are all the galaxies moving away from us at the same speed, or are the more distant galaxies moving away faster than the nearby galaxies?

- After doing some research, you might like to prepare models of various types of stars, such as white dwarfs, red giants, supergiants, pulsars, neutron stars, supernovas, quasars, and other stars.
- Can you make dynamic models of stars? Of a binary star system? Of a star's life cycle?

4
MODELS IN LIFE SCIENCE: FROM ORGANS TO GENES

In medicine, live animals are used to model human physiological responses to new drugs and a variety of medical treatments. One example is the guinea pig, which has been used so extensively that "serving as a guinea pig" has come to mean experimental involvement in a very broad sense. Medical researchers have also developed numerous models of human organs with the idea of constructing fully functional replacements for these organs. More than a decade ago artificial hearts were inserted within human chests. While their success was limited, researchers are optimistic about the prospect of fully functional artificial hearts, as well as artificial kidneys, lungs, and other organs. See Photo 16.

Models in biology have long served the purpose of helping researchers and students understand how complicated living systems work. These models range from actual physical models of biological structures such as DNA molecules,

Photo 16. Heart-lung and other machines
in use during heart surgery

chromosomes, and cells to complex mathematical models that describe population growth, heredity, and neurological activity in the brain.

In this chapter you'll have an opportunity to build some models that have a biological nature or flavor. While we can't offer any designs for truly functional artificial organs or perfect models for higher-order brain activity, perhaps someday you will!

HOW LUNGS WORK

A model often used to illustrate the way we draw air into our lungs is shown in Figure 15. You can make such a model quite easily. The rubber dam fixed to the bottom of the bottle represents the diaphragm that separates the chest and abdominal

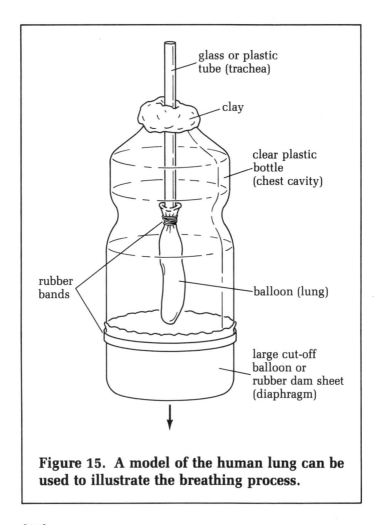

Figure 15. A model of the human lung can be used to illustrate the breathing process.

cavities. What happens to the balloon when you pull down on the "diaphragm"? Why does it happen? What happens when you release it? Why?

- The model you've built fails to take into account the role our rib cage plays in breathing and the fact that our lungs don't collapse every time our diaphragm muscle relaxes. See if you can build a more realistic working model of our lungs that better illustrates the breathing process.

ONE-WAY VALVES FOR BLOOD

Blood flows away from the heart through arteries and back to the heart through veins. Because the pressure of the blood in veins is very low, gravity would cause blood to fall to our legs and feet were it not for one-way valves in the veins that allow blood to flow only toward the heart.

You can see the effect of these valves quite easily. Let your arm hang at your side for a minute or so as you open and close your fist to increase blood flow to your hand and arm. You should now be able to see veins along the back of your hand or the inside of your forearm. Hold one finger firmly against the lower portion of a vein. Move another finger upward along the vein to sweep blood away from the region that you have closed off. You will find a region where the vein has collapsed. Just above the collapsed portion of the vein is where a valve is located. Blood can flow back down the vein as far as the valve, but no farther, because the valve allows blood to flow in only one direction—toward the heart.

Now see if you can build a model of a one-way valve like the kind found in veins. Build it in a piece of clear tubing so that you can see how it restricts the flow of a colored fluid, such as water and food coloring, to one direction. Normally, muscles contracting around the veins force blood toward the heart. In your model, you can squeeze the tubing to move the fluid along the tube.

OPEN YOUR HEART

If possible, obtain a sheep or cow heart. You can probably obtain one from a slaughterhouse, a butcher shop, or a scientific supply company. Either organ is very similar to the human heart.

Dissect the sheep or cow heart very carefully, *working under supervision of a science teacher*. You'll be able to see the valves that allow blood to flow in only one direction through the heart. From the veins of the body, the blood flows to the right auricle, or atrium, as you can see in Figure 16. When this auricle contracts, blood flows through the thin, leaflike tricuspid valve into the right ventricle. This chamber then pumps the blood forcefully through three pocketlike semilunar valves into the pulmonary artery—the artery that carries blood to the lungs, where it is oxygenated. Blood returning from the lungs via the pulmonary veins enters the left auricle, which contracts, forcing oxygen-rich blood through the bicuspid, or mitral, valve into the left ventricle. From there, the blood is forced through three aortic semilunar valves into the aorta—the main artery that carries blood to the body.

After dissecting an animal heart, see if you can build a working model of the human heart. You might use water mixed with red food coloring to represent blood. The real challenge will come in

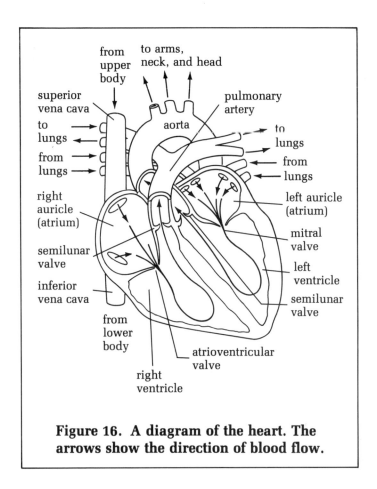

Figure 16. A diagram of the heart. The arrows show the direction of blood flow.

building a four-chambered organ with a pumping mechanism and valves that allow blood to flow in only one direction.

- Building functional artificial hearts is something that has been, and is being, researched extensively. You might like to consider some of the things that would have to be done to convert your model to an artificial heart that could be used to replace damaged human hearts.

SEEING IS BELIEVING

A model of the human eye is one of the most common projects found at science fairs. The model presented here requires you to do more than simply convert a textbook drawing into a three-dimensional model.

If possible, obtain a sheep or cow eye. You can probably obtain one from the same place where you found a heart. Working under supervision, dissect the eye carefully. Its structure is, again, very similar to that of the human eye. Based on what you have learned from your dissection, prepare a model of the human eye that will allow you to cast an image on the "retina" of your model eye. See if you can incorporate a lens that behaves like the one in your eye, that is, a lens that will change its shape so that images of both nearby and distant objects can be focused on the retina. Can you also include an iris with a pupil that will adjust its size depending on the intensity of the light?

KNEES AND OTHER JOINTS

Many of the joints in our bodies—the places where bones meet—enable us to bend and move our bodies in a great variety of ways. Our knees, elbows, and fingers are hingelike joints that allow us to flex (in one direction) and also to extend our legs, arms, and digits until they are straight. Our shoulders and hips are ball-and-socket joints that allow us to rotate our limbs. The pivotlike joint in your neck allows you to turn your head sideways as well as move it forward and backward.

You might enjoy building working models of some of the body's larger joints. Bear in mind, however, that the joints of the human body are seldom simple. The elbow, for example, which we generally think of as a hinged joint, has two bones on the lower side—the radius and ulna—that can rotate about one another as well as move in a hinged manner about the upper arm bone, or humerus. As you work on your models, you might get some ideas by reading about the artificial knee and hip joints that orthopedic surgeons are using to replace arthritic or otherwise seriously damaged joints. Scientists and doctors can now also construct computer-generated models of joints, which can then be used to assist in surgery. See Photo 17.

BUILDING BETTER MODELS

The biology laboratory in your school probably has models of DNA molecules arranged in a double helix, models of the changes in chromosomes that take place during mitosis (ordinary cell division) and during meiosis (the formation of egg and sperm cells), and models of the human eye, ear, teeth, and brain. Examine these models carefully.

Choose one or more of the models and see if you can devise a better and less expensive model using materials that you can find in your school or at home. When you have finished, ask your teacher if you may exhibit your model in the laboratory or in a hallway somewhere in the school.

A MATHEMATICAL SALAD (WITH PIZZA AND BURGERS)

Many of the fruits and vegetables you eat are peeled before you eat them. Others have a pit or core that you discard. How much of the fruit do

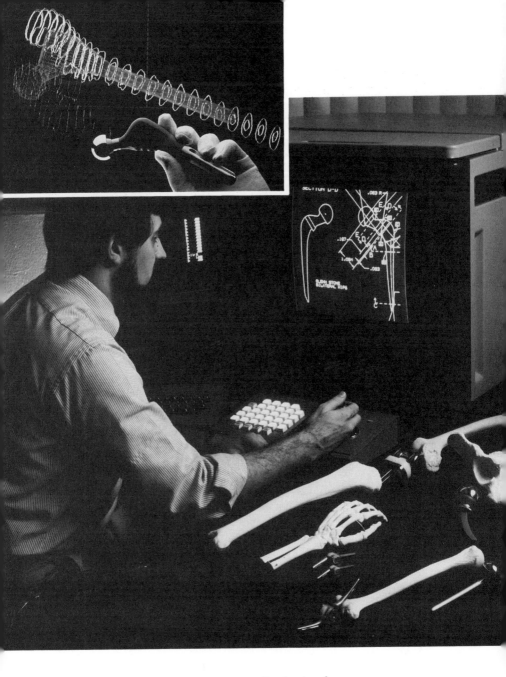

Photo 17. (a) Designing human joints on a computer. (b) (Inset) A computer-imaged human joint side by side with an artificial joint.

you eat and how much do you throw away? You can use a mathematical model and some approximations to answer this question.

You might start with an orange. It's reasonable to assume that it is a sphere; therefore, its volume is $4\pi r^3/3$, where r is the radius of the orange. Figure out a way to measure the diameter of the orange. Then calculate the volume of the orange. Now peel off the orange's thick skin, which you may assume has a uniform thickness. Measure the diameter of the peeled orange and calculate the volume of the edible part of the orange. From the measurements you have made, how can you calculate the volume of the orange's skin? From these measurements, calculate the thickness of the orange peel. (Remember, the surface area of a sphere is $4\pi r^2$.) Then measure, as accurately as you can, the thickness of the skin directly. How do the two determinations of the thickness compare?

Next, consider what fraction of a banana is edible. If you assume that it is a bent cylinder, its volume is $\pi r^2 L$, where r is the radius of the banana and L is its average length. Pick one banana from a bunch. What is the volume of the banana you picked? Now peel off the skin. What is the volume of the edible part of the banana? From the measurements you have made, how can you calculate the volume of the banana's skin? What do you calculate it to be? See if you can confirm this volume by measuring the thickness of the banana peel and determining its volume. You can assume the skin to be a rectangular solid. The width of the solid is $2\pi r$ (the circumference of the banana); its length is L (the average length of the banana); and its thickness is t, the thickness you measured. The volume of the banana peel, therefore, is $2\pi r L t$. How does this calculation of the banana peel's volume compare with the previous one?

How can you use a balance to determine what fraction of a banana or an orange is skin? What assumption are you making about the density of the banana and its skin if you use this method?

Now consider a pitted fruit such as a peach or a plum. What shape can you assume the fruit to be? The pit? Use mathematical modeling to find out what fraction of the total volume you eat and what fraction you throw away.

What fraction of an apple's volume do you discard if you eat it with the skin on? If you peel the apple before you eat it? How about a potato? A carrot?

What fraction of a pizza is crust? What fraction of a hamburger's volume is meat? What fraction of a hamburger's mass is meat? How about a hot dog? What fraction of a watermelon can you actually eat?

- Consider the fruits and vegetables from an economic standpoint. Taking into account the parts that are edible, which fruit provides the most food per dollar spent? The least?

KING KONG AND OTHER SCALING PROBLEMS

In an essay entitled "On Being the Right Size," the famous biologist J.B.S. Haldane wrote: "Consider a man sixty feet high—about the height of Giant Pope and Giant Pagan in the illustrated *Pilgrim's Progress* of my childhood. These monsters were not only ten times as high as Christian, but ten times as wide and ten times as thick, so that their total weight was a thousand times his, or about eighty to ninety tons. Unfortunately the cross sections of their bones were only a hundred times those of Christian, so that every square inch of giant bone had to support ten times the weight borne by a square inch of human bone. As the human thigh-bone breaks under about ten times the human weight, Pope and Pagan would have broken their thighs every time they took a step. This was doubtless why they were sitting down in the picture I remember. But it lessens one's respect for Christian and Jack the Giant Killer."

After poking fun at the impossibility of the characters in such fiction as *Pilgrim's Progress*, and he might have included *Gulliver's Travels* as well, Haldane points out that large animals are not simply blown-up versions of smaller animals that belong to the same biological order. A rhinoceros, for example, is much heavier than a gazelle, but to support so much more body weight its legs have become much thicker proportionally than those of the slender-legged gazelle. He goes on to say that a mouse, because it has so much surface area for its volume, "eats about one quarter its own weight of food every day, which is mainly used in keeping it warm. For the same reason small animals cannot live in cold countries."

Design models to show how length, surface area, and volume (or weight) are related to one another as objects are scaled upward and downward. Then show that because the strength of a bone or the supporting column for a structure is proportional to the area of cross section, this area should increase by the square root of 8 (2.82), not 2, when the scale of a structure is doubled. Show, too, why becoming wet, which we humans take in stride, is a serious problem for an insect, why hummingbirds must eat their own weight of food every day, and why the man-eating insects or King Kong–size gorillas of movieland could never exist in reality. What other factors related to length, surface area, and volume can you illustrate with your models of scaling?

- Based on what you've learned about scaling, you might like to devise a model to explain why some paleontologists believe the largest dinosaurs lived partly submerged in swampy areas. You may find some of the books in the bibliography helpful.
- Robert Bakker (see bibliography) has raised some questions regarding our current theories about dinosaurs. You might like to compare Bakker's model with the more traditional one and decide for yourself which, if either, is better. Perhaps you'll develop a model of your own to explain what we know about these extinct giants.

MODELING GENES

In the 1860s an Austrian monk and botanist, Gregor Mendel, investigated the way in which certain characteristics, or traits, found in garden peas were transmitted from one generation to the next. He began with true-breeding plants—plants in which the characteristics remain unchanged from generation to generation. He found that traits which disappeared in the first generation offspring often reappeared again in the second generation.

Mendel's experiments were ignored for nearly forty years until Hugo de Vries discovered his writings after carrying out similar experiments in 1900.

After you have read about the work of Mendel and deVries, see if you can build a visual model to explain their work. You might use pieces of yarn with different colors and lengths to represent chromosomes. Small boxes could represent the germ cells that give rise to egg and sperm cells, which in turn could be represented by even smaller boxes.

Your model should reflect the independent segregation of genes and the traits they produce. It should also reflect the role that chance plays in the combination of chromosomes and genes to form zygotes.

Not all genes show independent segregation. Under what condition will the traits produced by genes not segregate independently? What does your model lead you to predict about genes that are located on the same chromosome? (Such genes are said to be "linked.")

- During meiosis, chromosomes of the same pair sometimes cross over one another, break, and exchange segments—a phenomenon known as crossing-over. How can you modify your model to take crossing-over into account? What does your model tell you about the likelihood of genes becoming separated if they are close together on the same chromosome? If they are far apart on the same chromosome? What kinds of genetic experiments might be conducted to determine which genes are linked and how far apart they are on the chromosome?

5
MODELS IN CHEMISTRY: FROM CRYSTALS TO CHAIN REACTIONS

Chemistry is the scientific investigation of matter—its composition, structure, properties, and reactions. To understand the bubbles of gas that appear when a seltzer tablet is added to water, the beautiful crystals that form when a solution of alum evaporates, or the explosion of TNT, we must turn to the invisible world of atoms.

Chemical reactions, such as the ones mentioned above, result from the breaking and reforming of bonds between vast numbers of atoms. Since atoms are too small to be seen, chemists have used models to develop their understanding of them. In fact, the development and refinement of atomic models over the past two centuries represents one of the greatest achievements in science. These models have enabled chemists and physicists not only to gain a better understanding of matter and its reactions but also to predict the existence of previously undiscovered laws and the products of various chemical reactions.

DALTON'S ATOMIC THEORY

The first detailed atomic model was proposed by the English chemist John Dalton in 1803. It consisted of four postulates:

1. All elements are made of indestructible particles called atoms.

2. Atoms of the same element are identical to each other but different from those of any other element.

3. Atoms can bond together to form molecules, the basic particles that make up compounds.

4. Chemical reactions take place when atoms change their bonding arrangements.

Dalton's model became widely accepted because it explained many of the properties of matter known in his day. His first postulate—the indestructibility of atoms—explained the law of conservation of matter (matter cannot be created or destroyed). His second postulate—the idea of unique atom types—explained why elements cannot be broken down into simpler substances. The many possible combinations between the atoms of nearly a hundred naturally occurring elements to form molecules (Dalton's third postulate) explain why there are so many compounds. The combining or decomposing of substances to form new substances is explained by Dalton's fourth postulate.

Dalton's atomic model can also explain a fundamental chemical law, the *law of constant proportions,* which states that when elements combine to form a compound, they combine in a fixed ratio by weight. For example, when hydrogen burns in oxygen to form water, the weight ratio of oxygen to hydrogen is always 8:1. That is, 8 g of oxygen combine with 1 g of hydrogen to form 9 g of water, 80 g of oxygen combine with 10 g of hydrogen to form 90 g of water, 800 g of oxygen combine with 100 g of hydrogen to form 900 g of water, etc. The weight ratio is always the same regardless of the size of the sample.

Making Molecules

You can devise a concrete, visual model to show how Dalton's theory explains this fundamental chemical law by using a number of identical objects to represent the atoms of an element. Different objects can be used to represent different elements. For example, paper clips might be used to represent element C, while small steel washers and paper fasteners are used to represent elements W and F, respectively. Be sure that all the paper clips, washers, fasteners, or whatever you use are very nearly identical in weight and can be fastened together. After all, the properties of all samples of an element are always identical.

Now make several samples of the same "compound." For example, if you use paper clips and washers, you can make "molecules" of the compound CW by attaching one washer to each paper clip. From two large samples of the elements C and W, combine them in a "reaction" to make two or three separate samples (piles) of the compound CW. Then "decompose" one of the samples into its elements C and W. Weigh both elements. What is the ratio of the weight of element C to the weight of element W in this sample of CW? Repeat the experiment for each sample you have prepared.

What do you find about the weight ratio of C to W in each sample? How does your model illustrate the law of constant proportions?

You can also make more-complicated molecules. A molecule of C_2W, for example, would consist of two atoms (paper clips) of C attached to one atom of W (washer). Prepare several samples of C_2W. How do these samples illustrate the law of constant proportions? Now prepare several samples of each of the following compounds: C_2W_3, CF, CF_2, and CW_2F_3. Do samples of each of these compounds illustrate the law of constant proportions?

Your concrete model of atoms and molecules might lead you to another fundamental chemical law. This often happens in science. A model may suggest the existence of relationships that lead to new laws. What law might be suggested by using the measurements that you have made with such compounds as CF and CF_2 to compare the weight ratio of F to C in these two compounds? How about CW, C_2W, and C_2W_3? Can they be incorporated into the same law?

- Elements sometimes combine to form more than one compound. For example, hydrogen and oxygen combine to form hydrogen peroxide (H_2O_2) as well as water (H_2O), nitrogen and oxygen unite to form a variety of compounds—N_2O, NO, NO_2, N_2O_3, and N_2O_5—carbon and oxygen react to form both carbon dioxide (CO_2) and carbon monoxide (CO), etc.
- If, for example, the mass of hydrogen is kept constant, the ratio of the mass of oxygen in hydrogen peroxide to that in water is 2:1. In

the case of CO_2 and CO, there is twice as much oxygen per gram of carbon in CO_2 as there is in CO. For a fixed mass of nitrogen in the compounds mentioned, the mass of oxygen in the series is 1:2:4:3:5.

These compounds illustrate the *law of multiple proportions*, which states that if the same elements combine to form two or more different compounds, the ratio of the masses of the elements in the compounds are small whole numbers. Examine the data you collected when you "decomposed" samples of CF and CF_2, and CW, C_2W, and C_2W_3. Did your data lead you to develop the law of multiple proportions?

ISOTOPES: THE SAME AND DIFFERENT

Early in the twentieth century it was discovered that the atoms of many elements are not identical in weight. Hydrogen, for example, can be separated into two kinds of atoms. Most of them (99.98 percent) have a weight of 1 atomic mass unit (1.7 x 10^{-24} g or 0.0000000000000000000017 g), but the remaining 0.02 percent have a mass twice as large. Similarly, many other elements have atoms that differ in atomic mass but not in their chemical properties. Atoms of the same element that differ in mass but have the same chemical properties are called *isotopes*. Table 4 lists the isotopes of a few elements and their relative abundance. Notice that the most common form of an element—hydrogen, for example—is also called an isotope. Can you modify your concrete, visual, paper clip model of atoms to take into account the isotopic forms of the elements? In this model, how sensitive would

Table 4. Natural isotopes of some common elements

Element	Isotopes and their (atomic) weights	Relative abundance (%) of each isotope
hydrogen	1	99.98
	2	0.02
helium	3	0.01
	4	99.99
carbon	12	98.89
	13	1.11
chlorine	35	75.53
	37	24.47
iron	54	5.82
	56	91.66
	57	2.19
	58	0.33
uranium	234	0.01
	235	0.72
	238	99.27

your weighing device have to be in order to separate atoms or molecules that contain different isotopes of the same element?

THE RISE AND FALL OF THE PLUM PUDDING ATOM

The discovery of electrons by the English physicist J. J. Thomson in 1897 revealed that atoms were more complex than Dalton's model suggested. Electrons, which carry a fundamental unit of negative electric charge, were found to be a common building block of all atoms. However, because their weight is only one two-thousandth the weight of the lightest atom (hydrogen), their contribution to the total weight of any atom is very small.

As a result of this discovery, Thomson developed a new model of the atom, sometimes referred to as the plum pudding model. He proposed that the vast majority of an atom's mass resides in a jellylike clump of positive charge. The electrons, enough of them to exactly balance the positive charge, are distributed throughout this clump like raisins in a plum pudding.

In 1911, the British physicist Ernest Rutherford and two of his students designed an experiment to test Thomson's model. They directed a beam of high-energy alpha particles (helium atoms without their electrons) at a very thin sheet of gold foil, one that was only a few hundred atoms (10^{-5} cm) thick. They expected to see the alpha particles pass through the foil with very little deflection or velocity change because the "jellylike" atom with its tiny imbedded electrons was not expected to offer much resistance.

A sketch of their experiment is shown in Figure 17. You can imagine their surprise when they detected a few alpha particles that had been deflected through angles as large as 180°. Rutherford said: "It was quite the most incredible event that has ever happened to me in my life. It was almost as incredible as if you fired a 15-inch shell at a piece of tissue paper and it came back and hit you. . . . I realized that this scattering backward must be the result of a single collision and when I made calculations I saw that it was impossible to get anything of that order of magnitude unless you took a system in which the greater part of the mass

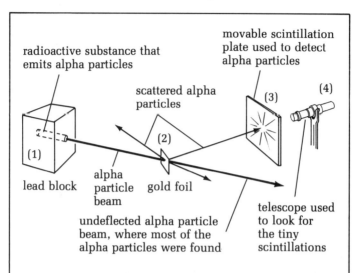

Figure 17. Rutherford used this apparatus to find out what happens when alpha particles are fired at a piece of thin gold foil. The entire apparatus was enclosed in a vacuum chamber. A radioactive substance (1), such as polonium, provided the alpha particles, which were fired at the gold foil (2). A movable plate (3) that emitted tiny flashes of light when struck by alpha particles was used to detect the deflected particles. The tiny scintillations could be seen only by using a telescope (4) to magnify the light.

of the atom was concentrated in a minute nucleus. It was then that I had the idea of an atom with a minute massive center carrying a [positive] charge."

By analyzing the number of deflected alpha particles and their angles of deflection, Rutherford was able to calculate that most of the atomic mass and all of its positive charge were concentrated in a very tiny region called the nucleus. For gold atoms the nucleus was less than 10^{-12} cm in radius—one ten-thousandth the total radius of the atom. You can see why positively charged alpha particles would be deflected if they came close to an atom's positively charged nucleus. But because most of an atom's space is empty, the probability for such a collision is small. Consequently, most of the alpha particles passed through the gold foil undeflected. Instead of confirming Thomson's model, Rutherford's experiment had led him to develop a new model of the atom.

Nailing Nuclei and Other Atomic Modeling

In one model illustrating Rutherford's scattering experiment based on his nuclear theory of the atom, widely spaced nails in a flat board represent gold nuclei. Small steel balls "fired" at right angles to the line of nails can serve as alpha particles. If you know the number of nails (N), the total width of the board (L), the number of hits per thousand particles "fired" (H), and the radius of the steel ball alpha particles (r), can you calculate the radius of the "nuclei" (R)?

This is a good exercise in probability and statistics. By dividing the total width of the lanes along which the fired balls will strike a nail by the total width of the line of the board, you should be able to derive the expression

$$H = \frac{2N\,(r + R)}{L} \times 1{,}000,$$

which, when solved for R, yields

$$R = \frac{LH}{2{,}000N} - r.$$

Another approach to this model might be to use small magnets to represent nuclei. Magnets mounted as pendulum bobs on long strings could represent alpha particles. What other models might you build to illustrate Rutherford's scattering experiment?

- How does the model of the atom developed by the Danish physicist Niels Bohr (1885–1962) compare with Rutherford's atom? Can you devise a concrete model of the Bohr atom? How about more modern models of the atom that incorporate the ideas of the French physicist Louis de Broglie (1892–1987) and quantum mechanics?
- Find out more about the process by which the image in Photo 18 was made.

IONS: POSITIVE AND NEGATIVE ATOMS

Although electrons are bound to atoms by the attraction of the positively charged nucleus that they surround, it is possible to remove one or more electrons from an atom by supplying energy to the atom. Once an electron is removed, the atom is left with a positive charge. Sometimes an atom acquires an extra electron and becomes negatively charged. In either case, the charged atom is referred to as an *ion*.

As charged ions, they have much stronger attractions to oppositely charged ions than do neutral atoms or molecules. Furthermore, the energy required to ionize the atoms of an element is a

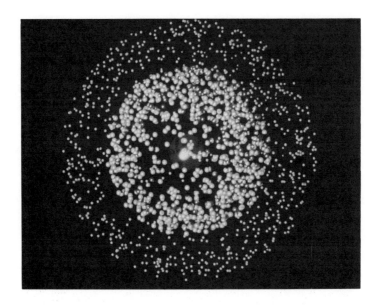

Photo 18. A computer-graphics representation of the electronic structure of a lithium atom

characteristic property that can be used to identify the element and to determine forces and energies within the atoms. Furthermore, the energy required to remove an electron from an element's atom provides valuable information about how readily the element will react with other elements and compounds.

The ionization energy is determined by bombarding the gaseous state of the element with electrons or photons (tiny particles of light) of known energy. When the energy of the bombarding particles reaches the precise level required to remove an electron from the atom, there will be a sudden surge of current due to the released electrons or the positive ions formed.

Modeling Ionization Energy

You can make a model to illustrate the ionization energy of an atom by using an inclined grooved plastic ruler to represent an atom, a marble at the base of the ruler to represent an electron within the atom, and a second marble on another identical, inclined ruler to represent a bombarding electron. See Figure 18. How can you vary the energy

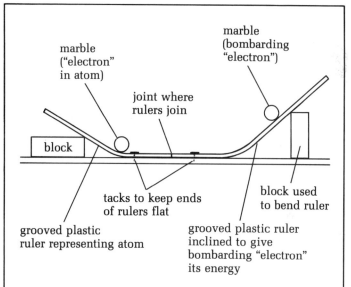

Figure 18. A marble is used to represent a bombarding electron. It acquires its energy by rolling down an inclined plane. When it collides with another marble, representing an electron bound to an atom, it transfers its energy to the bound electron. If the energy transferred is sufficient, the bound electron will "escape" (go over the ramp) and the atom will be ionized.

of the bombarding electron? How can you determine the energy needed to ionize the atom? How can you modify the apparatus to show that the ionization energy is different for different elements?

In this model, gravity is used to model the force that holds the electron to the atom's nucleus. How can you modify the model so that the force attracting the electron to the nucleus is an elastic force or an electric force?

RADIOACTIVE DECAY AND HALF-LIFE

Some elements have radioactive isotopes. The nuclei of these isotopes can be represented by a formula such as ^3_1H, an isotope of the element hydrogen. The subscript 1 preceding the H, which is the symbol for hydrogen, indicates the number of protons or positive charges within the nucleus. The superscript 3 indicates the mass number of the nucleus—in this case 3 (1 proton + 2 neutrons). The difference between the two numbers (3 − 1) gives the number of neutrons in the nucleus.

Radioactive isotopes release radiation in the form of alpha, beta, and gamma rays. In so doing, they become either less energetic nuclei or atoms of a different element. For example, the radioactive isotope sodium-24 ($^{24}_{11}Na$) releases a beta particle (an electron) from its nucleus to become $^{24}_{12}Mg$. Since the electron carried away one negative charge, the nucleus now has twelve positive charges, characteristic of magnesium isotopes, instead of the eleven characteristic of sodium isotopes.

Other nuclei release alpha particles ^4_2He (helium nuclei) to become isotopes whose nuclei have two less positive charges and four less atomic mass units. With such radioactive isotopes, a specific fraction of the element's atoms will decay, that is, release radiation in a fixed period of time, but it is impossible to predict exactly which atoms will decay.

Modeling Radioactive Decay with Sugar Cubes

This process can be modeled by throwing a large number of sugar cubes that have been marked on one of the six surfaces. For reasonable results, start with at least a hundred cubes. After each throw, some of the cubes (about one sixth if you have a large number) will land with the mark facing up. These cubes can represent decayed atoms, which are then removed before the next throw. Each throw of the cubes can represent a specific period of time—1 hour, for example. Continue the experiment until very few cubes remain.

Plot a graph of the number of "undecayed atoms" (the number of cubes remaining) versus time (hours if each roll represents 1 hour). What is the half-life of this "radioactive element"? That is, how long (how many throws) does it take before only half of the radioactive atoms remain? How many radioactive atoms remained after two half-lives? After three half-lives?

- Sometimes the radioactive atoms that have decayed undergo one or more additional decays. For example, uranium-238 releases an alpha particle in decaying to thorium-234. The half-life of the uranium-238 is 4.5 billion years. Thorium-234, which is a "daughter" element of uranium-238, has a half-life of 24 days and emits a beta particle (electron) to become protactinium-234. Can you modify the model of radioactive decay to show the decay of a daughter element with a different half-life than the parent?

A NUCLEAR CHAIN REACTION

In atomic reactors, uranium and plutonium atoms undergo fission. That is, the nuclei of these atoms split when struck by slow-moving neutrons to produce nuclei of smaller atoms. For example, when a neutron enters the nucleus of an atom of the uranium-235 isotope (see Table 4), the atom may fission to produce barium-141 and krypton-92 as well as three neutrons. Because these new nuclei and neutrons possess less energy than did the original single nucleus and neutron, the extra energy is released as kinetic energy in the products.

If the atoms are arranged so that one or more of the three neutrons released in the fission process causes another atom to fission, then a chain reaction will result. The neutrons released by the first atom to fission might cause two more atoms to fission; these two might each cause two more atoms to fission (a total of four); and so on. The fission process takes very little time, and each fission releases millions of times more energy than when equal numbers of atoms combine in ordinary chemical reactions. As a result, a fission chain reaction releases vast amounts of energy in a short time, producing an atomic (really a nuclear) explosion.

In atomic reactors the excess neutrons are absorbed by control rods. Only one of the neutrons released causes another atom to fission. As a result, the chain reaction can be controlled and energy is released at a steady rate.

A Chain Reaction with Dominoes

To make a model of a chain reaction, gather together as many dominoes as you can find. Arrange the dominoes so that if you push one of them it will topple two; these two will each topple two more, and so on. What does each fall of a domino represent? Design some other models of a chain reaction. Can you include some way of illustrating the energy release that accompanies the reaction?

In a fusion reaction, nuclei of hydrogen isotopes fuse to form helium nuclei. This reaction results in the release of even larger amounts of energy than occur in nuclear fission. Develop a model to illustrate a fusion reaction.

ATOMIC PACKING IN CRYSTALS

The existence of naturally occurring crystals, such as diamond and quartz, with their angled faces, suggested to early scientists that the atoms in crystals were arranged in ordered layers. Today scientists know this for certain. By recording on film the patterns made when X rays pass through thin crystal specimens and then analyzing them, scientists routinely identify the structure of crystals. Also, the recently invented scanning tunneling electron microscope can actually produce images of ordered rows of atoms that make up crystal surfaces. Photos 19, a and b, show such an image.

The identical atoms of metallic elements, such as iron, copper, and aluminum, form crystals

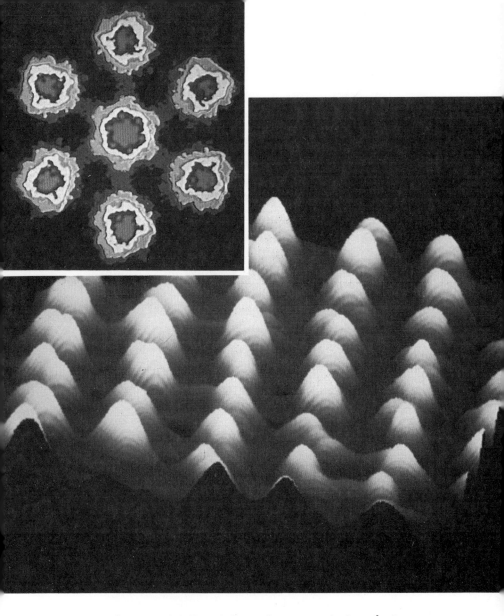

Photo 19. (a) (Inset) Scanning transmission electron micrograph of a crystal of a uranium compound. Each spot is an image of a single uranium atom. The images are magnified millions of times!

(b) Scanning tunneling electron micrograph of an organic substance. Each peak represents the image of an atom.

by packing together as closely as possible. To investigate these close-packed arrangements of atoms you'll first need several dozen identical Styrofoam balls to serve as models of metal atoms. These can be obtained from most crafts or fabric stores.

Start by building a single plane of close-packed atoms (connect them together with small drops of glue or toothpicks). You'll find that closest packing requires that each atom be surrounded by six others in the same plane as shown in Figure 19a. Call this first close-packed plane of atoms layer A. Next, build a second close-packed plane (layer B) and lay it on top of layer A, as shown in Figure 19b. You'll find that closest packing requires that the atoms of layer B fit into every other one of the "holes" formed at the junctures of three atoms in layer A.

Now build a third plane of close-packed atoms. There are two ways to place this third layer and still maintain closest packing. The first is to place it so that each new atom lies directly over an atom of the original layer A. The new atoms are placed over junctures marked X in Figure 19b. That is, you can repeat layer A. With this choice you could build a crystal structure based on the alternating close-packed layers ABABAB. . . . The second way is to place the atoms of the third layer over the alternative holes of the original layer. The new atoms are placed over junctures marked Y in Figure 19b. This would represent a new layer C. A close-packed crystal structure could then be based on the alternating close-packed layers ABCABC. . . .

Go ahead and build models representing these two close-packed crystal structures. Examine them from different angles and see if you can identify

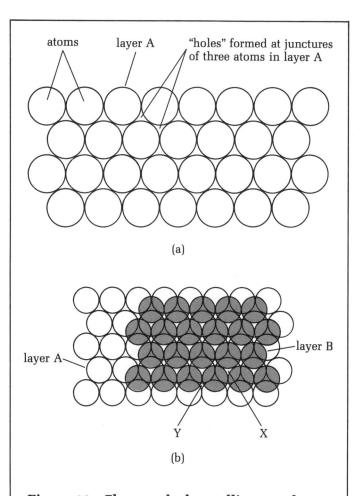

Figure 19. Close-packed metallic crystal structures. In (a) you see a layer of identical atoms. "Holes" are formed at the junctures of three atoms. In (b) you see a second layer of atoms over the first, arranged on top of the holes between the atoms in layer A. When a third layer of atoms is added, their centers may be placed over either the X holes or the Y holes.

differences. How many nearest (touching) atomic neighbors does each atom have?

Can you make other close-packed patterns?

IONIC CRYSTALS

When metallic elements such as sodium, potassium, calcium, and barium react with nonmetallic elements such as oxygen, fluorine, and chlorine, there is a strong tendency for the outer electron(s) of the metal atoms to transfer to the nonmetal atoms. In losing electrons the metal atoms become positively charged ions (called cations). Conversely, in accepting extra electrons the nonmetal atoms become negatively charged ions (called anions). The strong attraction that results between these oppositely charged ions leads to the formation of ionic crystals.

A common example of an ionic crystal is sodium chloride (NaCl), or table salt, which is made up of strongly bonded sodium cations and chlorine anions (see Photo 20). Common to all ionic crystal structures is the requirement that the immediate neighbors of each cation are all anions, and vice versa. This is a result of both the strong attraction between ions of opposite charge and the strong repulsion between ions of like charge.

Models of Ionic Crystals

To investigate how ions might pack together to form an ionic crystal separate some identical Styrofoam balls into two groups. One group will represent cations; the other, anions. Use paint or markings to distinguish the cations from the anions. Next, build a crystal model having equal numbers of cations and anions such that the near-

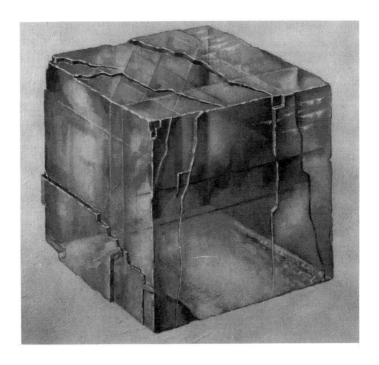

Photo 20. Salt crystal

est neighbors for each ion all have the opposite charge; that is, cations are surrounded by anions, and anions are surrounded by cations. Can this condition be met by each ion having twelve oppositely charged nearest neighbors? Is this ionic crystal structure close packed? How many oppositely charged nearest neighbors does each ion have?

The ionic crystal model you built represents the structure of ionic compounds such as potassium fluoride (KF) or cesium chloride (CsCl). These crystals are made up of equal numbers of cations and anions that are nearly equal in size (K^+ and F^- or Cs^+ and Cl^-). A drawing of one of these structures is shown in Figure 20. In many ionic

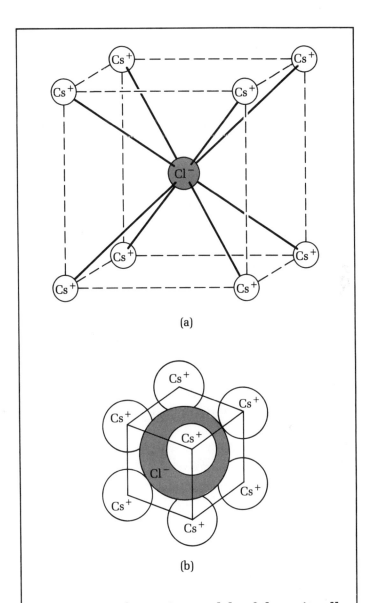

Figure 20. Alternative models of the unit cell of cesium chloride (CsCl). (a) Ball and stick model. (b) Spherical model.

compounds the cation and anion have different sizes and this influences the crystal structure. Generally speaking, the closer the ions are in size, the greater the number of close oppositely charged ion neighbors. The cation-to-anion size ratio is a factor that determines ionic crystal structure. Table 5 gives the relative size of a number of ions. Table 6 shows how the coordination number (the number of anions close to the cation) depends on the cation-to-anion size ratio.

Another factor that determines ionic crystal structure is the relative number of cations and anions in the compound. For example, calcium fluoride (CaF_2) has two fluorine anions for every one calcium cation. This means that every cation has twice as many anions surrounding it as there are cations surrounding each anion.

To illustrate how the ion size and number factors affect the ionic crystal structure, consider the examples of sodium chloride (NaCl) and thorium oxide (ThO_2). NaCl has a cation-to-anion ratio of 0.95/1.81 = 0.52, which, according to Table 6, means that each sodium cation is surrounded by six chlorine anions, and vice versa. This structure is shown in Figure 21. CsCl, on the other hand, has a cation to anion size ratio of 1.69/1.81 = 0.93, which means that each cesium cation is surrounded by eight chloride anions, as you have seen. In the case of thorium oxide (ThO_2), which has a borderline ratio of 0.71, the coordination is 8. Because there is just one thorium ion for every two oxide ions, each oxide anion has only four thorium ions as close neighbors.

The structure of ThO_2 is shown in Figure 22.

Table 5. Relative ionic radii of selected ions

Li^+	Be^{2+}		O^{2-}	F^-
0.60	0.31		1.40	1.36
Na^+	Mg^{2+}	Al^{3+}	S^{2-}	Cl^-
0.95	0.65	0.50	1.84	1.81
K^+	Ca^{2+}		Br^-	
1.33	0.99		1.95	
Rb^+	Sr^{+2}		I^-	
1.48	1.13		2.16	
Cs^+	Ba^{2+}			
1.69	1.35			
	Ce^{+4}			
	1.72			
	Th^{+4}			
	1.00			

Table 6. The effect of cation-to-anion ratio on crystal structure

Ratio of cation radius to anion radius	Arrangement or coordination of ions about central ion	Coordination number (number of close neighbors)	Diagram of arrangement of atoms in crystal
1-0.732	corners of cube	8	
0.732-0.414	corners of octahedron	6	
0.414-0.225	corners of tetrahedron	4	
0.225-0.155	corners of triangle	3	

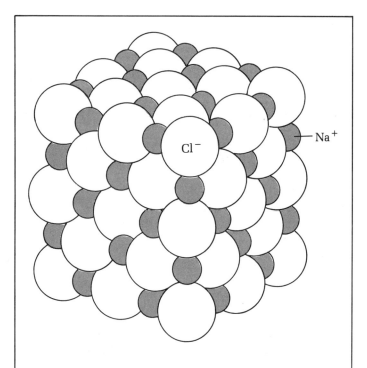

Figure 21. The crystal structure of sodium chloride. Each chloride ion is surrounded by six sodium ions; each sodium ion is surrounded by six chloride ions.

- Try building models of the following ionic crystal structures using Styrofoam balls of approximately the correct scale size: beryllium oxide (BeO), magnesium oxide (MgO), cesium bromide (CsBr), and cerium oxide (CeO$_2$). Refer to Table 5 for atomic radii and to Table 6 for coordination numbers and crystal structure.

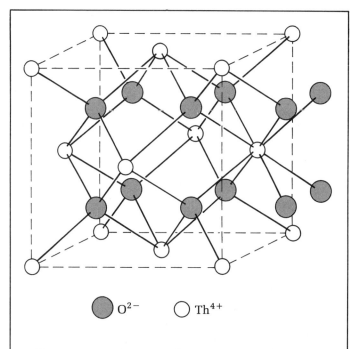

Figure 22. The crystal structure of thorium oxide (ThO$_2$). Each oxide ion has four thorium ions as close neighbors.

NOT ALL CRYSTALS ARE PERFECT

Perfect crystal structures do not exist. All naturally occurring crystals contain numerous microscopic flaws. Even the artificial crystals grown in sophisticated laboratories have numerous defects on the atomic level. For instance, some atoms will be missing from their normal positions (vacancies) or they may be forced into odd positions between normal positions (interstitials). Also, impurity atoms can replace normal atoms (substitutional impurities) or be forced into spaces between other atoms (interstitial impurities). These types of imperfections are called point defects.

Defects can also take the form of misplaced planes of atoms. For example, extra planes of atoms can be squeezed into an otherwise perfect lattice. The effect is similar to the way a large bookmark disturbs the well-ordered pages of a book. An even more disruptive type of defect occurs where the surfaces of two microscopic crystals (grains) having different orientations meet. These "grain boundaries" are similar to the boundaries formed between bubbles in a foam. Most naturally occurring crystalline solids exist as agglomerates of tiny grains ranging in size from one ten-thousandth of a centimeter to one one-hundredth of a centimeter. Large single-grain crystals are very rare, which is why diamonds are so valuable.

Modeling Imperfections

To make a two-dimensional model of crystal imperfections, cover half the bottom of a tray or shallow box with a single layer of metal shot. If you tilt the tray, the shot will collect on one side and form a model crystal structure. You should be able to identify regions of close-packed order (grains), grain boundaries, vacancies, and interstitials. You can also model impurity defects by introducing a few shot of a different color and size.

Try to form a perfect crystal without directly touching the shot. What factors can you control to improve the order of the crystal or increase or decrease the grain size and number of defects? Can you think of what real crystal variables correspond to these factors?

- Devise a way to test how defects might move (diffuse) under the action of external stresses.
- Use the model to investigate: (1) how substitutional and interstitial impurities might diffuse through the model crystal, (2) the factors that determine the rate of diffusion, (3) the role that vacancies play in diffusion.

6
MODELS IN THE PHYSICS OF MOTION: FROM FALLING BODIES TO SPEEDING BULLETS

The earliest civilizations used models to explain the motions of celestial and earthly objects. For example, the ancient Greeks invented the geocentric model of the universe in which the stars, planets, sun, and moon revolved around the stationary earth in perfect circular orbits. They believed that this type of motion was required by the "perfect" or divine nature of heavenly bodies. Earthly objects, on the other hand, were compelled by their "imperfect" nature to fall toward the center of the earth. Furthermore, earthly objects were said to fall at constant speed; however, heavy objects were believed to fall proportionately faster than lighter ones.

It was not until the fifteenth century that the great Italian astronomer and scientist Galileo Galilei (1564–1642) challenged this model. The story is told of how he simultaneously dropped two stones, one large and the other small, from the Tower of Pisa in full view of the medieval professors who still believed the ancient Greek model. The stones fell together and hit the ground almost simultaneously, demonstrating that the weight of the stones was of little significance in determining their rate of fall.

FREE-FALL: FASTER AND FASTER

Galileo was not satisfied with simply showing that heavy and light objects fell at the same rate. He also wanted to show that their speed increased uniformly with time. This type of motion, called uniform acceleration, can be expressed mathematically by the equation

$$v = at,$$

where v is the speed of the falling body, t is time, and a is the constant acceleration. (In this equation it is assumed that the falling object starts from rest; that is, $v = 0$ at $t = 0$.) While Galileo had no means of measuring speeds directly, he could measure distances and times. The touchstone of his genius was his ability to derive the distance-time relationship that constant acceleration leads to. This relationship is

$$d = \tfrac{1}{2} at^2,$$

where d is the distance fallen. This meant that he could demonstrate constant acceleration indirectly by showing that the distance a body falls increases with the square of the time. Actually, Galileo used geometric arguments rather than algebra.

Galileo faced one practical difficulty in measuring free-fall motion. Because free-falling bodies rapidly attain speed, they cover large distances in only fractions of a second. Timing devices that could measure such small times were not available to Galileo. To solve this problem, Galileo developed a physical model of free-fall. He proposed that a ball rolling down a shallow incline would exhibit the same pattern of motion as a freely falling body. The only difference would be that the speeds would be reduced proportionately. In other words, the effect of the incline would be to uniformly "dilute" the effect of gravity.

Testing Galileo's Model

You can build your own version of Galileo's inclined plane model of free-fall and use it to test his mathematical model of free-fall. Start with a straight, rigid, 3-m section of aluminum or steel angle iron. Lay it on a floor with the V facing upward. Elevate one end about 15 cm with pieces of wood or books. This produces an angle of incline of about 3°. Mark off 20-cm intervals starting from the elevated end. A hard rubber or steel ball can serve as the "falling body."

With a stopwatch, measure the times it takes the ball to roll from rest at the starting point (0 distance) to each marked distance along the incline. When you release the ball, be careful not to give it any backspin. This can probably best be achieved by placing a pencil across the front of the ball. At the moment you wish to release the ball, pull the pencil forward, parallel to the metal angle.

Record the results in a table and then plot the individual distance-time data pairs on a distance-versus-time graph. Do the distance-time data fit the relationship expressed by $d = \frac{1}{2} at^2$? One way to check is to calculate the ratio $2d/t^2$ at each point and see if it is constant. Since $d = \frac{1}{2} at^2$, $2d/t^2$ is the acceleration. Another approach would be to plot a graph of the distances versus the times squared. What should this graph look like if acceleration is constant? What will be the slope of such a graph?

Repeat the experiment with steeper inclines. Are the accelerations constant for steeper inclines? What happens to the acceleration as the incline becomes steeper? Develop a data table of acceleration versus angle of slope and graph the results. Can you extrapolate the graph to obtain a value for the acceleration at a slope of 90°? Can you think of reasons why this extrapolation might not be perfectly valid?

Modern timing devices are capable of measuring very small time intervals and therefore can directly determine the distance-time relationship for freely falling bodies. These measurements show that small, dense objects do fall with constant acceleration over a few meters. The actual value of the acceleration depends slightly on the object's location. This is because the pull of gravity varies slightly from place to place on the surface of the earth. At the equator, the acceleration is about 9.8 m/s². These precise timing devices reveal that air resistance can cause falling bodies to deviate from constant acceleration. You might be aware that air resistance increases with the speed of the falling object and with the air pressure. At high speeds, the resistant force of the air can be as great as the force of gravity. When this occurs, the body no longer accelerates downward but reaches its *terminal velocity*.

Another Model of Constant Acceleration

Even without precise timing devices, you can still demonstrate the constant acceleration of free-fall with the following model. Cut a 260-cm piece of thread and tie five steel bolts along its length as shown in Figure 23. Notice that the distance from the bottom bolt to the others increases according to the sequence of factors: 1^2, 2^2, 3^2, 4^2, 5^2. This is the same sequence of factors by which the free-fall distances increase with time (i.e., $d = \frac{1}{2} at^2$).

Therefore, if the bolts are suspended vertically with the bottom bolt just resting on the ground and the remaining bolts suspended at the indicated heights, then, when released, the bolts would fall and strike the ground at equal time intervals. Although you cannot measure these time intervals, by listening carefully to the thuds when the bolts hit the floor you can tell that the time intervals are very nearly equal. To make the sounds more distinct, let the bolts fall on a metal tray.

CANNONBALL IN MOTION

Galileo, who showed the world that objects fall with a constant acceleration, also discovered that the horizontal and vertical motions of projectiles, such as balls launched by cannons (or today by the arms of outfielders and shot-putters), are independent of each other. If a bullet is fired horizontally from a gun at the exact moment that another is dropped from beside the gun, both will strike the ground at the same time. The bullet that falls straight downward will land directly below its release point. The bullet fired horizontally will move with a constant horizontal velocity while falling

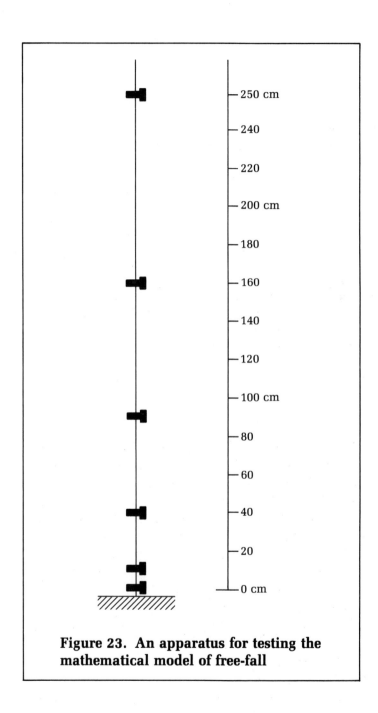

Figure 23. An apparatus for testing the mathematical model of free-fall

vertically with the same acceleration as the ball that was dropped. Both will accelerate downward at 9.8 m/s², which means that their downward velocity increases at the rate of 9.8 m/s every second.

A Model of Projectile Motion

You can make a simple model to illustrate this principle. With your hand turned downward, gently hold two marbles side by side between your thumb and fingers. Use a finger on your other hand to snap one of the marbles horizontally outward. Once the marbles separate, the stationary marble will start its fall to the floor while the marble you struck continues to move horizontally and downward. Listen carefully! How many sounds do you hear as the marbles strike the floor?

To make a more extensive and convincing model, build a ramp using a grooved plastic ruler, a flat board, tacks, and a wood block, as shown in Figure 24. Clamp the board in place and use the ramp to launch a steel ball or marble projectile. Hang a plumb line, such as a paper clip on a thread, from the end of the ramp. The end of the paper clip will mark a point directly beneath the launch point of the projectile.

Let the steel ball or marble roll down the ramp about a dozen times, always from the same height on the ramp. A sheet of white paper taped to the floor and covered with carbon paper will cause the projectile to leave marks when it lands. If the ball or marble is released from the same point each time, the marks on the paper should be quite close together.

Measure the distance from the lower end of the plumb line to the center of the marks made by the projectile. What horizontal distance does the projectile travel during its flight to the floor?

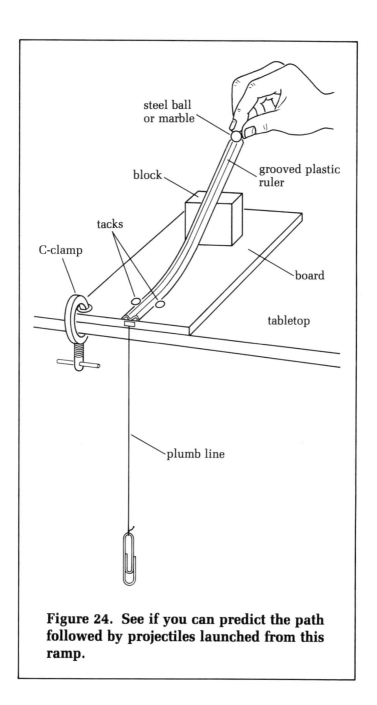

Figure 24. See if you can predict the path followed by projectiles launched from this ramp.

To find the horizontal velocity of the ball, you need to know how long it took the ball to reach the floor. From Galileo's free-fall model, we can easily find the time. Since $d = \frac{1}{2}at^2$, the height, h that the projectile falls in any time, t, is given by

$$h = \tfrac{1}{2} at^2.$$

You can find the height by measuring the distance from the bottom of the projectile at the end of the ramp to the floor. Since the acceleration is 9.8 m/s², you can calculate the time it takes the ball to reach the floor because

$$t^2 = \frac{2h}{a}, \text{ and so } t = \sqrt{\frac{2h}{a}}.$$

Suppose the projectile falls 1.0 m during its flight to the floor. Then the time it takes for it to reach the floor is

$$t = \sqrt{\frac{2h}{a}} = \sqrt{\frac{2 \times 1.0 \text{ m}}{9.8 \text{ m/s}^2}} = 0.45 \text{ s}.$$

If the horizontal distance, d, that the ball traveled during its flight is 0.50 m, its horizontal velocity was

$$v = \frac{d}{t} = \frac{0.50 \text{ m}}{0.45 \text{ s}} = 1.1 \text{ m/s}.$$

Once you know the projectile's time of flight and the horizontal distance it traveled, you're ready to make a chart and build the model. The chart should resemble Table 7. It indicates the horizontal and vertical positions of the projectile at

Table 7. Horizontal and vertical distances for a projectile launched in a horizontal direction

Time (s)	Horizontal distance traveled (m)	Vertical distance traveled (m)
0.05	0.055	0.012
0.10	0.11	0.049
0.15	0.17	0.11
0.20	0.22	0.20
0.25	0.28	0.31
0.30	0.33	0.44
0.35	0.39	0.60
0.40	0.44	0.78
0.45	0.50	1.23

various times during its path to the floor. The chart shown here is based on the data from the examples above where the time of flight was 0.45 s and the horizontal velocity was 1.1 m/s. Your data will depend on the way you build the "launching pad." Once you have prepared your chart, use it to make a model of the projectile's path. Small (½-inch) Styrofoam balls—the "snowballs" used to decorate Christmas trees or to make molecular models—or corks can be threaded and hung from a horizontal rail clamped beside, and extending out from, the projectile launching ramp as shown in Figure 25. If the top of the rail is at the same height as the end of the ramp, vertical distances, taken from the chart, can be measured from the top of the rail. Horizontal distances, also taken from the chart, can be measured from the end of the ramp.

To test your model, simply launch a projectile as you did before. Does its path match the path of the model you have made?

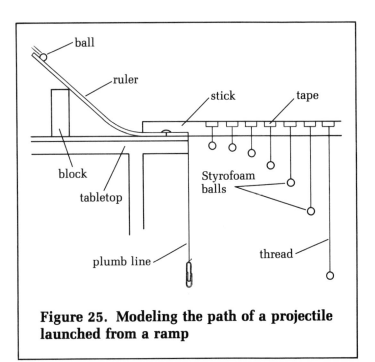

Figure 25. Modeling the path of a projectile launched from a ramp

- Prepare a similar model for a projectile launched upward or downward at an angle.
- In the model you've made here, you assumed air resistance to be negligible. You were justified because you used relatively heavy objects that never reached large speeds. But suppose you had fired a bullet, such as the one mentioned at the beginning of this section, which travels at a high velocity. How would you take air resistance into acount? Or suppose you used an object such as a leaf or a sheet of paper, which is so light for its surface area that it stops accelerating very quickly because of air resistance and thereafter falls with a steady, rather slow terminal velocity.

- Design a model to illustrate how the air resistance on an object is affected by its velocity, surface area, density, and shape.
- How can a model of projectile motion be used to predict the landing point of a marble launched from a grooved ruler? Of a baseball thrown toward home plate by an outfielder? Of a long-range shell launched from a cannon? In which case would air resistance play the most important role?

7
MODELS IN THE PHYSICS OF LIGHT: FROM MIRRORS TO SLINKIES

For those of us who can see, much of what we know about the world comes to us through our sense of sight, through the light that reaches our eyes. The light entering our eyes is brought together to form images by the lenses within our eyes. Sometimes other lenses—those in magnifying glasses, microscopes, telescopes, or spectacles—are placed in front of our eyes to enhance the images we see. Sometimes that light is brought to our eyes by mirrors, which reflect light.

Sir Isaac Newton, who did so much to simplify and bring understanding to the complexities of motion, was also the major force in unraveling the mysteries of light. It was Newton who not only thoroughly explored the properties of light, but also proposed a unifying model to explain it as well. However, the Dutch physicist Christiaan Huygens (1629–1695), a contemporary of Newton, offered an alternative model. Newton argued convincingly that light consists of tiny fast-moving particles; Huygens, on the other hand, believed that light could better be explained by thinking of it as wavelike. Both theories can be used to explain the way light reflects from common flat mirrors or the more complex curved ones found in giant telescopes, such as the one at Mount Palomar.

Regardless of what light is, we can use the scientific laws associated with light—the laws of reflection and refraction—to explain how a great variety of optical devices work. Building two-dimensional models of mirrors and lenses will help you to understand how they work. Using two dimensions instead of three makes the models simpler. After you've built the two-dimensional models, you'll be able to extend what you have learned to the real world of three dimensions.

The simplicity of the models can also be enhanced by restricting the light you are viewing to a few key, narrow beams of light that we'll refer to as light rays. In the abstract model of optics a ray has no dimension other than length; it is infinitely thin. In your models, the rays will not be ideal. If they were, you wouldn't be able to see them. You'll have to compromise and let a very narrow beam of light represent a ray.

PLANE MIRROR IMAGES

If you look directly into an ordinary plane (flat-surfaced) mirror, you see an image of yourself. Light traveling from you to the mirror is reflected back to your eyes. Your image appears to be as far behind the mirror as you are in front of it. When you approach the mirror from in front, your image approaches the mirror from behind. See Photo 21.

To understand why the images seen in plane

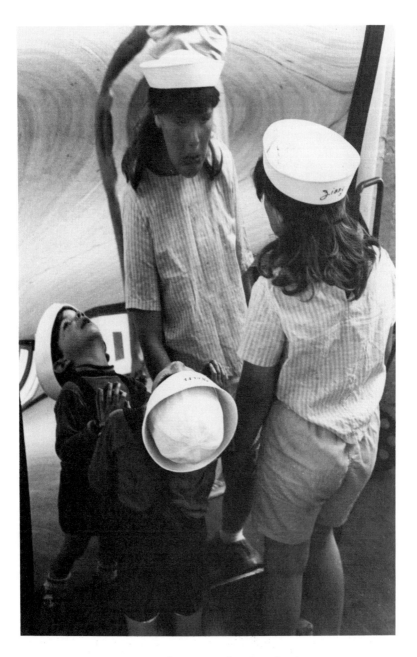

Photo 21. Playing with curved mirrors in the fun house at Coney Island

mirrors on a wall appear to be behind or imbedded in the walls, you can make a model to show what happens to light that strikes a plane mirror. To make the model as simple as possible, you can limit the light to a single point that emits just two rays. By using rays of light, we can examine what happens to a very small part of the light and assume that similar things will happen to the rest of it.

You can build a ray maker as shown in Figure 26. Cut away one side of a box. In the opposite side, near the bottom, cut an opening about 8 cm high and 10 cm wide. From a sheet of heavy, black construction paper, cut a mask about 10 cm by 12 cm that you can use to cover the opening in the box. Before you tape the mask to the box, use sharp scissors to cut two vertical slits about 8 cm high and 1 mm wide near the middle of the paper. The distance between the two slits should be about 1 cm.

The two narrow slits will allow you to make two rays of light when you illuminate them with a light bulb placed on the opposite, open side of the box as shown in Figure 26. A clear bulb with a straight-line filament works best. You can buy these in most supermarkets or hardware stores. Turn the bulb so that the filament is vertical (parallel to the narrow slits in the mask). Take a few moments to move and turn the light about until you obtain the sharpest rays possible.

Use your ray maker to produce two sharply defined rays. A plane mirror supported by a small lump of clay can be used to cross the rays and make a point source of light as shown in Figure 27. A second mirror can be used to reflect the two rays coming from the point source. Look at the two rays reflected by this second mirror. As you can see, the

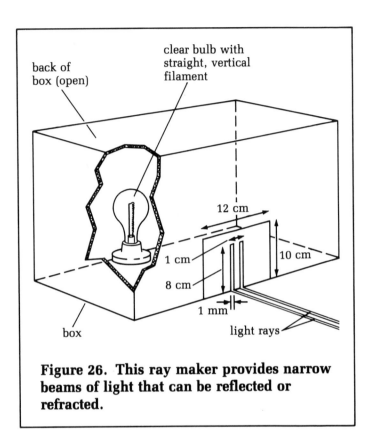

Figure 26. This ray maker provides narrow beams of light that can be reflected or refracted.

rays are diverging. If, in your imagination, you project these reflected rays backward behind the mirror to their apparent point of origin, where does that point appear to be? How does its distance behind the mirror compare with the distance between the mirror and the point source in front of the mirror? How does your model help you to understand why the image of an object appears as far behind a mirror as the object is in front?

- By experiment, you can show that the angle between a mirror and an incoming (incident) ray is equal to the angle between the

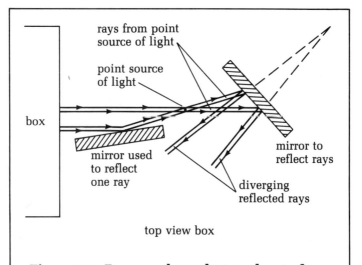

Figure 27. Rays are brought together to form a point source of light. Rays from this point can be reflected from a mirror.

mirror and its reflected ray leaving the mirror. (See Figure 28.) Using that information, develop a geometric model to show that the image of an object is as far behind a plane mirror as the object is in front.
* Suppose that an object is to one side of, rather than directly in front of, a plane mirror hanging on a wall. Will the image of the object still appear to be behind the wall? Use your model to find out. Extend your geometric model to explain what you find with your model.

CURVED MIRROR IMAGES

You'll find mirrors with curved surfaces on the sides of cars and trucks and in their headlights;

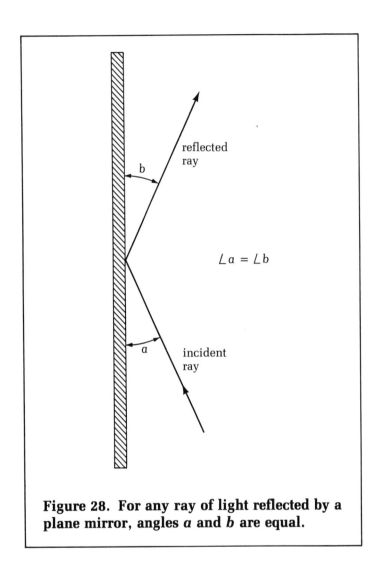

Figure 28. For any ray of light reflected by a plane mirror, angles *a* and *b* are equal.

you'll find them in stores, in the form of shaving and makeup mirrors, and you'll find them in such scientific instruments as telescopes. A *convex* mirror has a reflecting surface that bulges slightly outward, like an inverted saucer. The surface of a *concave* mirror, on the other hand, bends inward like a saucer or the inside of a bowl. A polished soup spoon can serve as a rough kind of mirror too. The side that holds the soup is like a concave mirror. The other side is similar to a convex mirror.

If you look into a convex mirror like the ones found on the sides of trucks and cars and in stores where they are used to spot shoplifters, you'll notice that the images, like those in a plane mirror, are behind the mirror. However, the images are smaller and appear farther away than do the images seen in plane mirrors. Furthermore, the field of view seen in a convex mirror is broader than the one seen in a plane mirror.

Concave mirrors are called converging mirrors because they bring light rays together. The images seen in a concave mirror depend on the position of the objects. You can make models of both kinds of curved mirrors.

Two-Dimensional Models of Curved Mirrors

The ray maker you used in the previous model can be used to establish rays that can be reflected from curved surfaces. In a dark room place the two-slit mask you used before over the opening in the box. Let the two rays strike a concave cylindrical mirror or something similar, such as a section of a cylindrical box lined with smooth aluminum foil. Notice how the two reflected rays are converged (brought together) by the concave mirror.

Now use a plane mirror to change the path of one of the rays coming from the light box. Deflect (reflect) the ray so that its path is parallel to that of the other ray. Place the cylindrical concave mirror so that it reflects the parallel rays. The point where the reflected parallel rays come together is the focal point of the mirror. The distance from the center of the mirror to the focal point is the focal length of the mirror. What is the focal length of your concave cylindrical mirror?

Now modify your model so that you can see what happens to rays of light that come from two different points on an "object." You'll need a mask with two pairs of slits. The slits in each pair should be about 1 cm apart, and the distance between the right-hand member of one pair and the left-hand member of the other pair should be about 1.5 cm. Place different-colored cellophane or plastic over each pair of slits so that the rays can be identified. Use a pair of plane mirrors, as shown in Figure 29, to cross the rays. The two points where the rays of the same color cross can be thought of as the top

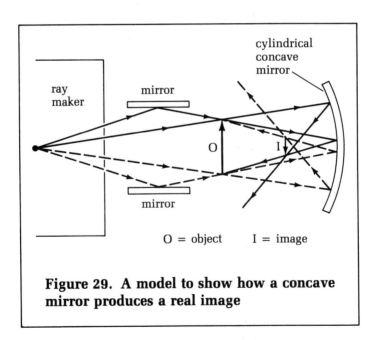

Figure 29. A model to show how a concave mirror produces a real image

and bottom of an "object," shown as an arrow marked O in the drawing. A concave cylindrical mirror can be used to reflect the rays. The points where the two colored rays of each pair of rays come together again mark the positions of the top and bottom of the image.

What does your model tell you about the image? Is it right side up or upside down as compared with the object? How is this image different from the one seen in a plane mirror?

By moving the cylindrical mirror, the distance between the object and the mirror can be changed. What happens to the position and size of the image when the mirror and object are brought closer together? When they are moved farther apart?

What happens when the distance between the mirror and the object is less than a focal length? Use your imagination to extend the reflected rays. Where is the image now? How do you know?

If you turn the mirror around, you will have a convex mirror. Where is the image of the object now? What happens to the image as the mirror moves closer to the object? Farther from the object? Do you see why convex mirrors are used as side mirrors on trucks and in stores to detect shoplifters?

Using Models to Make Predictions

Your work with two-dimensional models of concave and convex mirrors should enable you to make predictions about a three-dimensional concave mirror. If possible, obtain such a mirror. You may be able to borrow one from your school's science laboratory, or you can use a makeup or shaving mirror.

To find the focal length of the concave mirror,

you can use the light coming from the sun or any distant object. **If you use the sun, don't look at it! It can damage your eyes.** The light rays from distant objects are very nearly parallel. Therefore, the place where the rays are brought back together by the mirror will be the focal point. (Remember, the focal point is where reflected parallel rays meet.) If the sun is shining, you can hold a cardboard sheet in front of the mirror and find the point where the sun's image is sharpest. If you stand on the far side of a room with a window through which you can see clouds or a distant mountain or building, you can produce a sharp image of the distant object on a card held in front of the mirror. Measure the distance from the mirror to the sharply defined image. What is the focal length of your mirror?

Based on what you saw using your two-dimensional model of a concave mirror, predict what you will see if you hold the real concave mirror close to your face—that is, less than a focal length away. Were you right?

Now use the model again to predict what you will find if you look at the image of an object placed more than one focal length from the mirror. How will the image compare with the object? How will its size and location change as the object is moved closer to, or farther from the mirror? You can use a small clear light bulb, such as a flashlight bulb, as the object. Light the bulb and place it in front of the concave mirror at a distance greater than the mirror's focal length. Move a file card back and forth in front of the mirror until you locate the bulb's image. Is the image upside down or right side up? Does the image change size and position as you predicted when you move the light farther away from the focal point of the mirror?

Move the light so it is close to the mirror—less than a focal length away. Where is the image now? How can you locate the image?

The inverted images that you found in front of the concave mirror are called real images because they really consist of an infinite series of points where light rays emitted from the object are brought back together again. As you've seen, you can actually capture such images on a screen. The upright images that appear to be located behind mirrors are called virtual images. They can't be captured on a screen; the diverging rays that give rise to such images appear to have their origin behind the mirror.

Based again on your two-dimensional models, what kinds of images would you expect to find using a convex mirror? How will the position and size of the image depend on the position of the object?

Obtain a real convex mirror and test your predictions. (You can always use the sideview mirrors on trucks or cars.) Were your predictions correct?

- Devise a model to show how a reflecting telescope works.
- Devise a model to show how a periscope works.

A RAY MODEL OF A CONVEX LENS

Lenses are optical devices that can bend (refract) light. They can converge light rays to form real images, like the images formed by concave mirrors when they reflect light from objects outside their focal lengths, or they can diverge light as convex mirrors do when they reflect light.

The same ray maker that you used with mirrors can be used to model the formation of images with a convex lens. A clear glass or plastic jar filled with water can serve as the "lens." Begin, as you did before, with two rays. Place the lens on the rays. What happens?

To find the focal length of your cylindrical lens you will need a pair of parallel rays. How can you use the parallel rays to find the focal length of your lens?

As before, mirrors can be used to cross two pairs of colored rays emerging from the light box. The points where the like-colored rays cross represent the top and bottom of an object. If the lens is placed more than a focal length from the "object," it can be used to bend the rays and form an image as shown in Figure 30. Are the rays converged or diverged by the lens. Is the image upside down or right side up? Is it real or virtual?

By moving the lens, the distance between the

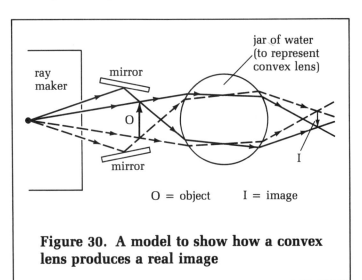

Figure 30. A model to show how a convex lens produces a real image

lens and the object can be changed. What happens to the size and location of the images as the lens is moved farther from the object?

What happens when the lens is moved to a point less than one focal length from the object? What happens to the rays that have passed through the lens? Do they converge or diverge? What kind of image do you find now? Where is the image?

TESTING MODELS USING REAL LENSES

If you have a real convex lens (a hand lens will do), you can use what you have found with your model lens to predict the size, shape, and nature of the images formed when an object is placed at different distances from the real lens.

Begin by finding the focal length of your real convex lens. How can you do that? Then test to see if you can use your model to predict what will happen to image size and position when an object, such as a lighted flashlight bulb, is moved from far away to a point just beyond the focal point of the lens. Do your findings agree with the predictions you made using the model? What happens when the object is less than 1 focal length from the lens? Where is the image now? Is it real or virtual? Is it right side up or upside down? Do your results agree with the prediction you made based on your model?

- Make a two-dimensional model of a concave lens. Then use the model to predict the kinds of images produced by a real concave lens. How would the size and position of the images seen through such a lens depend on the position of the object? Finally, obtain

a concave lens and test the predictions you made based on your model.
* Make a model to show how concave and convex lenses can be used to help people who are farsighted or nearsighted.
* Design a model to show how a microscope works.
* Design a model to show how a refracting telescope works.

LIGHT: WAVES OR PARTICLES?

A ray of light that strikes water, glass, or any other transparent material at an angle bends (refracts) sharply as it passes from one substance to another. You saw this in your model of a convex lens. Newton and Huygens both developed models to explain refraction as well as the other properties of light. According to Newton, the tiny fast-moving particles that constitute light are attracted toward glass, water, and other media only when they are very close to such surfaces. Thus, the fast-moving particles of light change direction only when they meet such surfaces. They change direction because one medium, such as water, pulls more strongly on the particles than another, such as air. Once inside a uniform material, they move in a straight line because they are pulled equally in all directions.

Huygens, on the other hand, believed that light is wavelike and that its behavior can be explained by considering it to be made up of periodic waves. He noted that waves, such as sound waves, travel faster in some media than in others. In order to change speed, the waves would have to bend. Since the speed of a water wave depends on the depth of the water, the waves bend when they pass from deep water into shallow water.

These two models, or theories, of refraction are abstract. We cannot see particles of light nor can we see light waver. Newton invented the idea of particles to explain the way light behaved. Similarly, Huygens imagined light moving like waves and argued that light's behavior could be explained by assuming that light was wavelike.

Does Light Behave More Like Marbles or a Slinky?

See if you can develop analogies, that is, models made from material objects, that illustrate how light would be reflected and refracted if it consisted of tiny particles. You might use marbles, tennis balls, ball bearings, or whatever seems appropriate to represent particles of light.

Then develop another model to explain reflection and refraction in a way that Huygens might have done to illustrate his wave model. You could use waves on a clothesline or a Slinky, or water waves in a sink, puddle, or ripple tank to represent waves of light.

Which model do you think better explains the reflection of light? Why do you think so? Can the same model explain the refraction of light? Can you invent yet a third model to explain the behavior of light?

CLOSER IS BRIGHTER—WHY?

Any photographer will tell you that the intensity of light—the amount of light per unit area—decreases as you move farther away from a light source. In fact, when the distance from a point source of light doubles, the intensity of the light, as measured with a light meter, becomes one quarter as large. If the distance triples, the intensity is reduced to one ninth its initial value. This phenomenon, known as an inverse square relationship, can be expressed as

$$I = \frac{K}{d^2},$$

where I is the intensity of the light, K is a constant equal to the strength of the light source, and d is the distance from the light.

See if you can devise an abstract model—a theory—to explain this relationship. Then devise a model or analogy that uses real objects to illustrate your hypothetical model. You might use a fine mist of colored water droplets forced from an atomizer, droplets of paint from a spray gun, or light itself emitted from a very tiny (point) source. How does the area covered by the particles or light change as you go farther from the source?

- If light consists of particles or tiny bundles of energy called photons, what kind of model would best illustrate this inverse square relationship?
- If light is wavelike in nature, what kind of model would best illustrate this inverse square relationship?
- If you're familiar with computers, you might develop a computer program that would simulate a model designed to explain the inverse square relationship. Can you develop other computer programs that would

illustrate reflection and refraction of light?
- When light passes through glass, plastic, water, or other transparent filters, the intensity of the light is reduced. Devise a hypothetical model to explain this decrease in intensity. Then invent an analogue that illustrates the workings of your model.

WHAT CAN MAKE A LIGHT BEAM SPREAD?

Obtain a clear incandescent bulb with a straight-line filament. Turn the bulb so that the filament is aligned vertically. Then look at the glowing filament through a narrow slit such as that between two rulers held close together. Look at it also with your eyes almost closed so that the light has to pass through the narrow slits between your lashes. Notice how the light spreads out into a series of bright and dark bands.

The way light spreads out as it passes through narrow slits is called diffraction. Develop a theory of your own to explain diffraction. Then develop an analogue to show how your theory works.

8 MODELS IN ENGINEERING: FROM BRIDGES TO AIRPLANES

Engineers are the men and women who put the principles of science to practical use. Their work can be seen throughout the modern world, in the cars, planes, and trains we ride in; the bridges we cross; the skyscrapers we live and work in; and the electrical and electronic devices we use in our homes, offices, and schools. To develop and maintain these marvels, engineers have relied increasingly on sophisticated models. While relying heavily on mathematical models in designing materials that can be manufactured effectively and efficiently, they often turn to scale models and analogues when the structures under investigation are too sophisticated or complex to be handled by mathematics alone.

In the activities that follow, you'll work with models and ideas associated with large structures such as bridges, with aeronautics, and with simple mathematical models used in obtaining the most effective structure for a given use.

THE BOX PROBLEM

An open box can be constructed from a single square piece of cardboard by cutting squares from each of its corners and then turning up its sides as shown in Figure 31. How would you go about making a box with the greatest possible volume from a given piece of cardboard? A trial-and-error approach would be unacceptable because a wrong first guess at the corner size would destroy the only piece of cardboard given. However, with a mathematical model you can predict the optimum dimensions before cutting anything.

Let L represent the side length of the cardboard and h the side length of the square corner cuts. The volume V of the box would then be given by the equation

$$V = h(L - 2h)^2,$$

where $(L - 2h)$ represents the side length of the bottom of the box after the sides, with a height h, have been folded up. Notice that h can take on any value between 0 and $L/2$. Since these values lead to a volume of 0, the optimum value of h must lie between these extremes.

Consider a particular case in which $L = 12$ cm. To find the value of h that produces the box with the greatest volume, substitute a number of different h values into the volume equation and plot the results on a V-versus-h graph. If you are familiar with computer programming, you may want to do this by using a computer. By mathematical trial and error you can find the graph's maximum value, or peak. This peak represents the solution to the problem. Next, construct a few

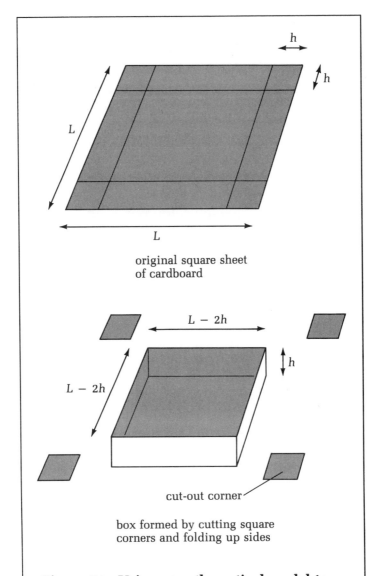

Figure 31. Using a mathematical model to obtain a box with the maximum volume from a given area of cardboard

boxes. Give one the optimum dimensions and the others slightly differing dimensions. By comparing the amounts of sand each hold you can prove to yourself that the mathematical model works.

- As an additional exercise, determine the maximum rectangular area that can be enclosed by a fence of a fixed length, say, 100 m. You might also try to find the dimensions of a closed can (i.e., its radius and height) that give the maximum volume for a given amount of sheet metal (total area of top, bottom, and sides). From your results, determine whether the manufacturers of canned foods and soft drinks are getting the most volume for the area of sheet metal used. If they are not, can you think of reasons why they may prefer not to use the optimum volume dimensions?

STRESS AND STRENGTH IN LARGE STRUCTURES

Large engineering structures such as bridges and skyscrapers require careful planning before money, materials, and labor are committed to their construction. This planning almost always involves extensive use of scale models to test and develop designs. These scale models go beyond mere similarity in appearance. They also provide similarity of function. That is, they allow the designers to accurately predict the strength, stability, and longevity of the structure before it is built. These models must also help the designers choose the most economic and sound materials and building methods.

Some structural modeling takes place on the materials level. For instance, new metal alloys are developed in small batches and their strength and corrosion properties are tested on small specimens. Often, model specimens are tested under extreme conditions, and statistical (mathematical) models are used to extrapolate their behavior for real-life conditions. Tests are also conducted on full-size structural components before they are incorporated into an actual bridge or skyscraper. Huge testing machines subject steel and concrete columns to enormous loads to determine their strength.

You probably have already spent a good deal of time modeling engineering structures without realizing it. By playing with blocks or toy construction sets you have actually tested a variety of basic designs that achieve maximum strength, size, and stability with the fewest parts. In the following activities you will continue building such models, but perhaps with a more conscious and directed effort.

Five Kinds of Stress

Before you start building complex structural models it is important that you learn a few guiding principles. First among these is the nature of stresses that exist in structural components that support loads. Stress arises when two or more forces are exerted simultaneously on different parts of a structure. There are five basic types of stresses (see Figure 32):

1. *Simple compression* arises from forces that push in on a structure through its center of mass. Top-loaded vertical support columns experience pure compression.

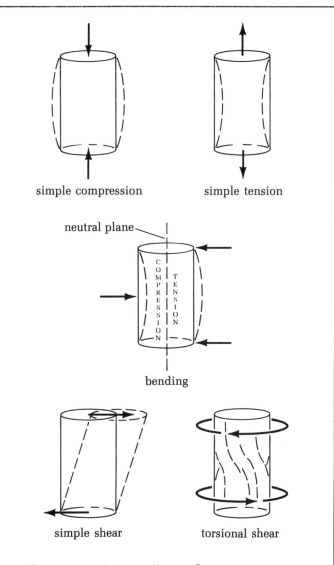

Figure 32. Stress arises when two or more forces acting on a body tend to change its shape. The diagrams show the five types of stresses.

2. *Simple tension* arises from forces that pull outward from opposite ends of a structure. Cables that support hanging loads experience pure tension.

3. *Bending stress* is a combination of tension and compression in the interior of a beam. The regions of tension and compression are separated by an imaginary "neutral" plane. A toy Slinky illustrates this behavior. When a Slinky is bent downward, its top half is in tension (the coils spread out) while the bottom is in compression (the coils are pushed together).

4. *Simple shear stress* arises when one part of a body is pushed sideways relative to another part. A spreading deck of cards illustrates shear stress. The pair of forces involved come from the hand that applies the sideways push along the top of the deck, and from the table, which holds the bottom of the deck by friction and prevents it from moving.

5. *Torsional shear stress* arises from the application of twisting forces. A twist-off bottle cap is opened by the application of a force that results in torsional shear stress.

Models of Stress

You might find it interesting and fun to make models that illustrate these five types of stress and their effects. One way is to start with five new bars of Plasticine clay and carefully apply a different type of stress to each with your hands. When they just begin to deform, remove the stress. Each type of stress will deform a bar in a unique way. You might cut out small cardboard arrows to represent the forces and attach them to each bar in their appropriate configuration.

Shaping Up for Strength and Stability

Another important factor in building structures is the role of geometric shape and orientation. To investigate a few of these factors, carry out the following simple activity.

Use a cardboard strip (2 cm by 10 cm) as a model of a simple beam. Use it to form a simple bridge by laying it flat across two books separated by 8 cm. Measure its support strength by loading it with coins or other small weights. Now try to improve this basic bridge design by reorienting, bending, or folding the beam. Were you able to improve the strength? Did you have to make certain trade-offs between strength, stability, and length of the bridge's span?

You probably discovered that the beam can carry a much greater load when it stands up on its edge. However, this orientation is unstable and requires side supports to remain upright. Other ways to improve the beam's strength are to fold it along its length, roll it into a cylinder, and bend it into an arch. The cylinder shape is most useful as a supporting column. In the case of the arch, end supports are needed to prevent its ends from slipping outward and flattening under a load. Also, arching reduces the effective length of the beam.

Rivets and Welds

In large structures many beams are joined by rivets or solid welds to create long spans. The strength and stability of these compound structures is determined by a few basic geometric principles. By carrying out the following activity you can discover these for yourself.

Join four cardboard strips end to end into a square structure by inserting paper fasteners through small holes punched in their ends (Figure 33). These fasteners serve as models for rivets. Next, join three other strips in a similar manner to form a triangle. Compare the stability of the square

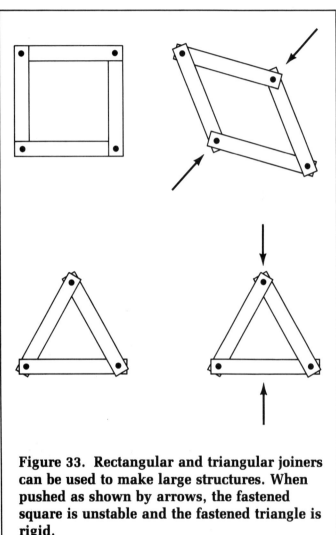

Figure 33. Rectangular and triangular joiners can be used to make large structures. When pushed as shown by arrows, the fastened square is unstable and the fastened triangle is rigid.

with that of the triangle. Try to analyze the nature of the stresses in the beams and joints of these structures.

Repeat this exercise by making a square and triangle with glued joints. (Glue can serve as a model for welds.) Again, compare the strength and stability of the two simple structures and try to determine the nature of the stresses in each beam and joint.

Which beams are in compression? Which are in tension? You can tell a beam is in tension if it can be replaced by a string (this is the idea behind suspension bridges). Which joints are subject to simple shear stress? Which are subject to torsional stress?

You probably found that the "riveted" triangle remained rigid under tension, compression, and shear forces. However, the "riveted" square structure collapsed under small loads. You should also have found that "welding" made the four-sided structure stiff. However, the tendency for the joints in the square structure to twist remains. This means that while the welded joints in the triangle structure must withstand simple shear stresses, those in the four-sided structure must also withstand both shear and torsional stresses. This makes the latter much more likely to fail.

BUILDING MODELS OF BRIDGES, TOWERS, AND SKYSCRAPERS

Using strips of cardboard, paper fasteners, string, and glue, try building some model bridges, towers, and skyscrapers. Incorporate the basic principles you learned above. See if you can analyze the different types of stresses that arise in the members and use this information to guide your design. Remember the following rules: (1) torsional stress should be avoided; (2) compressive stresses are best supported by folded beams or rolled columns; (3) pure tension stress in solid beams can be replaced with string; triangle components are inherently rigid.

Examples of some of the basic bridge types are shown in Figure 34. Notice how they incorporate the basic principles of strength and stability. You might try to incorporate some of these features into your own designs, but don't be afraid to try something new. You can learn as much from failed designs as from successful ones. After all, these are models and not actual bridges that people will use. There is no danger in experimenting with new ideas!

You might also have fun organizing a bridge-building competition among your friends and classmates. Set some guidelines with regard to the width to be spanned, the minimum strength requirements, and the amount and type of materials that can be used. You might judge the model bridges in different categories such as strength, appearance, and uniqueness of design.

LIFT, DRAG, AND OTHER EFFECTS OF FLOWING AIR

Aeronautical engineering is concerned with the design and construction of aircraft of all kinds. As such, it involves the development of body and wing shapes, internal support structures, propulsion systems, control systems, and basic materials. All these aspects must be combined to produce safe and reliable aircraft that will operate effectively and safely.

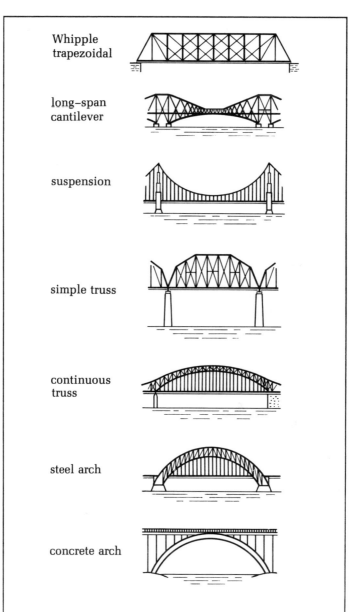

Figure 34. Some of the basic types of bridges

The great complexity involved in meeting such flight demands as speed, range, maneuverability, and payload size requires the use of extensive and sophisticated models. To test body and wing designs, engineers use scale models in conjunction with wind tunnels. In wind tunnel experiments the behavior of a plane flying through still air is simulated by blowing high-velocity air past a stationary model plane. Aeronautical engineers have achieved excellent similitude in such tests. As a result, aircraft designs can be tested safely, efficiently, and economically.

Wind tunnel experiments can determine such things as aerodynamic lift, drag, stability, frictional heating of surfaces, and the flow patterns of the air. Of course, wind tunnel experiments are also used in nonaeronautical applications. For example, most automobile designs are tested in wind tunnels to determine their aerodynamic drag. The effects of moving air on bridges and buildings are often tested in wind tunnels as well. Photo 22 shows a wind tunnel project at a science fair.

A Model Wind Tunnel

You can build a small wind tunnel fairly easily using a box fan and several cardboard boxes as shown in Figure 35. The tapered section acts both to increase the velocity of the air and make it more uniform across the face of the exit. To measure the airspeed at different points across the exit face, you can build a simple airflow velocity meter like the one in Figure 36. To calibrate the airflow velocity meter, mark the deflection of the cardboard vane when you walk at different speeds in still air. You can determine each speed by simultaneously timing how long it takes you to cover 20 m. Be sure to hold the meter level as you calibrate it.

Photo 22. Wind tunnel project at the 1992
Boulder Valley School District Science Fair

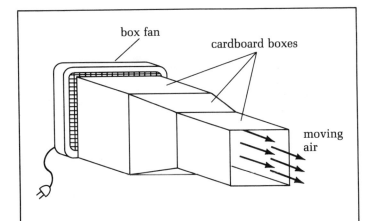

Figure 35. A homemade wind tunnel can be used to examine the effects of moving air on various types of wings.

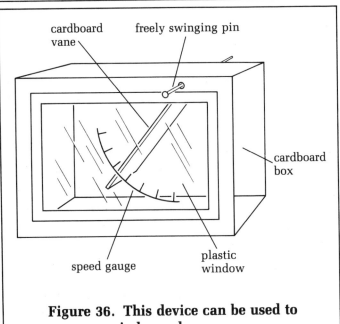

Figure 36. This device can be used to measure wind speed.

If the wind speed varies across the exit face of the wind tunnel, try stacking paper tubes such as paper towel tubes across the entrance to the last box. The tubes can be taped together and fastened to the sides of the box. You might also secure a sheet of window screening in the tunnel about 30 cm in front of the fan if variations in wind speed persist.

To investigate how aerodynamic drag depends on shape, build several different shapes out of Styrofoam or rolled newspaper wrapped in masking tape. Each shape should have the same frontal area. Build a stand consisting of a stiff but flexible wire set upright in a wooden block (Figure 37). Stick each shape on the wire and then place both in the path of the wind. By comparing the deflection of the wire, you can get a relative measure of the drag force. See if you can discover which shape produces the least drag. Does the shape of the tail that faces downwind have any effect on the drag?

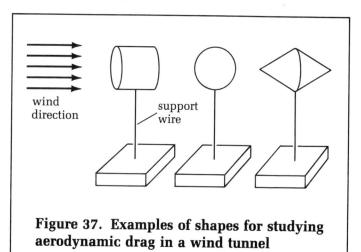

Figure 37. Examples of shapes for studying aerodynamic drag in a wind tunnel

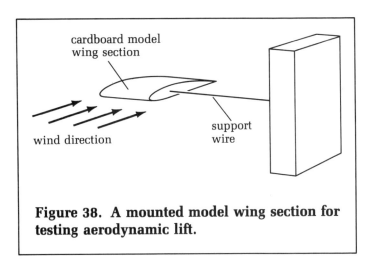

Figure 38. A mounted model wing section for testing aerodynamic lift.

Use the wind tunnel to test different wing designs by comparing their lift characteristics. You can build the wings from cardboard and mount them on a stiff wire as shown in Figure 38. Again, the relative lift force can be determined from the deflection of the wire. You might also build wings with control flaps and investigate how their position affects lift and drag. You might also test store-bought models and try to study the effects of modifying them.

By now you have made and used a variety of models. We trust that you have developed an understanding of the role that models play in science and engineering. The skills that you have acquired in designing and building these models are valuable and will serve you well in your own future researches, be they scientific or otherwise. For building and refining models, after all, is how humans grow in their understanding of the world around them.

BIBLIOGRAPHY

BOOKS

Apfel, Necia H. *Astronomy and Planetology: Projects for Young Scientists.* New York: Franklin Watts, 1983.

Bakker, Robert T. *The Dinosaur Heresies.* New York: Zebra, 1988.

Beller, Joel. *So You Want to Do a Science Project!* New York: Arco, 1982.

Caket, Colin. *Model a Monster: Making Dinosaurs from Everyday Materials.* New York: Sterling, 1979.

Challand, Helen J. *Activities in the Earth Sciences.* Chicago: Children's Press, 1982.

Dunbar, Robert E. *The Heart and Circulatory System: Projects for Young Scientists.* New York: Franklin Watts, 1984.

Gardner, Robert. *Energy Projects for Young Scientists.* New York: Franklin Watts, 1987.

———. *Experimenting with Light.* New York: Franklin Watts, 1991.

———. *Ideas for Science Projects.* New York: Franklin Watts, 1986.

———. *More Ideas for Science Projects.* New York: Franklin Watts, 1989.

Gardner, Robert, and Webster, David. *Science in Your Backyard.* New York, Messner, 1987.

Gillette, David D. and Lockley, Martin, eds. *Dinosaur Tracks and Traces.* New York: Cambridge University Press, 1991.

Lockley, Martin. *Tracking Dinosaurs: A New Look at an Ancient World.* New York: Cambridge University Press, 1991.

McKay, David W., and Smith, Bruce G. *Space Science Projects for Young Scientists.* New York: Franklin Watts, 1986.

Tocci, Salvatore. *How to Do a Science Fair Project.* New York: Franklin Watts, 1986.

Weiss, Harvey. *Model Buildings and How to Make Them.* New York: Harper Collins, 1979.

Zubrowski, Bernie. *Clocks: Building and Experimenting with Model Timepieces.* New York: Morrow, 1988.

ARTICLES

The following magazine articles provide information about artificial joints:

Cholakis, J. "Breakthrough: the synthetic knee." *Sport,* January 1988, p. 17.

Clark, M. "Engineering new hips." *Newsweek,* April 12, 1982, p. 78.

Sonstegard, D. A., et al. "Rumblings from the body shop: developing an artificial hip joint; spherocentris prosthesis." *Scientific American,* January 1978, pp. 44–51.

INDEX

Aerodynamics, 12
Aeronautical engineering, 148, 150
Aircraft, 148, 150
Analogues, 8
Anions, 99, 100
Aquifer, 27, 28–29, 31
Astronomical unit, 43
Astronomy, 36–65
Atoms, 8, 15, 80–84, 89–92

Bakker, Robert, 78
Big Bang theory, 64
Black holes, 37, 64
Blood flow, 69–70
Blue light, 33–34
Blue moons, 35
Blue skies, 33–35
Bode, Johann, 45–46
Bode's law, 45–46

Bohr, Niels, 89
Bridges, 147–148

Caloric theory, 12–13
Cassini, Giovanni, 47
Cations, 99, 100
Characteristics, of models, 13, 15
Chemistry, 80–107
Computer generated models, 6, 9
Computers, 8, 12
Contour line, 23–24
  and water table, 27
Copernicus, Nicholas, 43
Crick, Francis, 16
Crossing-over, 79
Crystals
  atomic packing, 95–99
  ionic, 99–104

Crystals (cont.)
    structure, 105–107

Dalton, John, 81
Dalton's atomic theory, 81–84
de Broglie, Louis, 89
Desert ecosystem, 18
de Vries, Hugo, 79
Diffraction, 137
Dinosaurs, 78
DNA molecule, 16, 17, 66, 73
Drag, 148, 150, 153, 154

Earth, 21, 45
    data, 46, 63
    measuring, 52–53
    to scale, 55–56
    view from space, 59–61
Earth science, 21–35
Eclipse, 56–58
Einstein, 64
Electrons, 15, 85–86, 89, 92
Elevation, 22, 23
Engineering, 138–154
    box problem, 139–141
Eratosthenes, 52–53
Eye, model, 72

Fabric of space, 64
Filtration, 29
Fission, 94–95
Flowing air, 148–154

Free-fall, 109–112
Frictional heating, 150

Galileo, 108–110, 112
Genes, 78–79

Haldane, J.B.S., 76–77
Half-life, 92–94
Heart, 66, 67, 69, 70–71
Heat, 12–13
Hereditary traits, 16
Huygens, Christiaan, 120–121, 134–135

Inventiveness, 15
Inverse square relationship, 136
Ionic crystals, 99–104
Ions, 89–92, 103
Isotopes, 84–85, 92–93

Joints, 72–73, 74
Jupiter, 45, 46

Kepler, Johann, 46–47
Kepler's third law, 47
Kidneys, 66
Kinetic energy, 94
Knees, 72–73

Law of conservation of matter, 81
Law of constant proportions, 81–82
Law of reflection, 121
Law of refraction, 121

Lenses, 120
    convex, 131–134
Life science, 66–79
Lift, 150, 154
Light, 33–35, 120–137
    intensity, 135–137
    particles, 134–135
    waves, 134–135
Lungs, 66, 68–69

Maps, 22–24
Mars, 45, 46, 47
Mathematical model, 8–12, 67
Medical treatments, 66
Meiosis, 73, 79
Mendel, Gregor, 78–79
Mercury, 45, 46
Milky way, 65
Mirror images
    concave, 126–131
    convex, 126–127, 129, 131
    curved, 122, 125–131
    plane, 121–125, 128, 129
Moon, 35
    diameter, 51–52
    distance to, 38, 40–41
    eclipsing, 56–58
    modeling, 38
    phases of, 37–38
    to scale, 55–56
Mount Everest, 63
Mount McKinley, 63

Neptune, 44, 46
Newton, Isaac, 120–121, 134–135
Nuclear chain reaction, 94–95
Nuclear theory of the atom, 88

Observation, 15

Physics of light, 120–137
Physics of motion, 108–119
Planets, 16
    distance to, 47, 49
    orbit, 45–46
Plate tectonics, 21–22
Plum pudding atom, 85–89
Pluto, 44, 46
Projectile motion, 114–119

Quantum mechanics, 89

Radar waves, 47
Radioactive decay, 8, 92–94
Radioactive isotopes, 92
Rain, 27
Rainbows, 31–33
Red light, 33
Red sunsets, 33–35
Relativity, 64

Richer, Jean, 47
Rivets, 145–147
Rutherford, Ernest, 86–88, 89

Safety, 20
Salt crystal, 100
Satellite data, 63
Saturn, 45, 46
Scale models, 12
   practical use, 7–8
Scaling problems, 76–78
Science and science fairs, 7–20
Science fairs, 7–20
   models, 16–17
Similarity in behavior, 8
Similitude, 13
Skyscrapers, 147–148
Soil erosion, 24–26
Solar car project, 19
Solar system, 61–63
Speed of light, 47
Stability, 145, 150
Star trails, 44
Stream formation, 25–26
Strength, 145
Stress, 141–144
Sun, 33–34
   damage to eye, 37, 40, 49, 55, 56, 130
   diameter of, 49–50
   distance to, 38, 40–41, 43, 47, 49
   eclipsing, 56–58

Terminal velocity, 111
Theory, 12
Thomson, J.J., 85–86
Three-dimensional scale, 24
Topographic mapping, 22–24
Towers, 147–148

Uniform acceleration, 109
United States Geological Survey, 24
Universe
   expanding, 63–65
   new model, 43–45
   two-dimensional model, 65
Uranus, 44, 46

Valves, 69–70
Venus, 45, 46, 47

Water, 27–30
   cleaning, 29–30
   table, 27–29
Watson, James, 16
Weather patterns, 12
Welds, 145–147
Wilkins, Maurice, 16
Wind tunnels, 12, 14, 16, 150–154

Zygotes, 79

L.A MATHESON SECONDARY
LIBRARY